Operation of Water Resource Recovery Facilities Study Guide

2018

Water Environment Federation
601 Wythe Street
Alexandria, VA 22314-1994 USA
www.wef.org

About WEF

The Water Environment Federation (WEF) is a not-for-profit technical and educational organization of 33,000 individual members and 75 affiliated Member Associations representing water quality professionals around the world. Since 1928, WEF and its members have protected public health and the environment. As a global water sector leader, our mission is to connect water professionals, enrich the expertise of water professionals, increase the awareness of the impact and value of water, and provide a platform for water sector innovation. To learn more, visit www.wef.org.

Prepared by the **Operation of Water Resource Recovery Facilities Study Guide** Task Force of the **Water Environment Federation**

Under the Direction of the **Technical Practice Committee**

Water Environment Federation Technical Practice Committee Control Group

Eric Rothstein, C.P.A., *Chair*
D. Medina, *Vice-Chair*
Jeanette Brown, P.E., BCEE, D. WRE, F.WEF, *Past Chair*

H. Azam
G. Baldwin
Katherine (Kati) Y. Bell, Ph.D., P.E., BCEE
J. Davis
C. DeBarbadillo
S. Fitzgerald
T. Gellner
S. Gluck
M. Hines
B. Jones
R. Lagrange
J. Loudon
C. Maher
S. Metzler
J. Miller
C. Peot
R. Pope
R. Porter
L. Pugh
J. Reeves
A. Salveson
S. Schwartz
A. Schwerman
Andrew R. Shaw, P.E.
A. Tangirala
R. Tsuchihashi
N. Wheatley

CONTENTS

ACKNOWLEDGMENTS

The following individuals contributed to reviewing and updating this new edition of the *Operation of Water Resource Recovery Facilities Study Guide*:

Bulbul Ahmed, Ph.D, P.E.
Jaime Alba, P.E., CRL
Jeanette A. Brown, P.E., BCEE
Natalie Cook
Bruce L. Cooley, P.E.
Viraj deSilva, Ph.D., P.E., BCEE
Richard Finger
Georgine Grissop, P.E., BCEE
Rhonda Harris
Natasha Jansen, EIT
Murthy Kasi, Ph.D., P.E.
Salil Kharkar, P.E., BCEE
Isaiah LaRue
David M. Mason
Nashita Naureen
Yue Xu, Ph.D.

We also wish to acknowledge the Task Force that developed the 2007 edition upon which this edition is based.

Jeanette A. Brown, P.E., BCEE, *Chair*

Merton Alexander, P.E.
Charles G. Farley
Richard E. Finger
Paul E. Fitzgibbons, Ph.D.
Michael T. Fritschi, WWTPO Grade V
W. James Gellner, P.E.
Frank Godin, MMP
Samuel S. Jeyanayagam, Ph.D., P.E., BCEE
Bruce M. Johnston
Lorene Lindsay
Jorj A. Long
Jon Meyer
Sharon M. Miller, P.E.
Robert Moser, P.E.
Gary W. Neun

Stephen Nutt, M.Eng., P.Eng.
Lokesh Padhye, MSCE, EIT
Joseph C. Reichenberger, P.E., BCEE
Melanie Rettie, PMP
Reza Shamskhorzani, Ph.D.
Robert B. Stallings, P.E.
Michael W. Sweeney, Ph.D., P.E.
Holly Tryon
David L. Ubert
Ifetayo Venner, P.E.
Milind V. Wable, Ph.D., P.E., BCEE
Jason D. Wert, P.E., BCEE

Chapter 1
INTRODUCTION

This study guide is a companion to the seventh edition of the Water Environment Federation's *Operation of Water Resource Recovery Facilities* (Manual of Practice No. 11; 2016). The goal is for these two publications to be principal training documents for facility managers, superintendents, and operators of municipal water resource recovery facilities as well as college students and consulting engineers. The manual and study guide can be used for training classes, preparing for certification exams, and improving the quality of operations within the water resource recovery facility (WRRF) or firm. As with the updated manual, this study guide reflects the state of the art in facility management and operation. The questions emphasize principles of treatment, facility management, troubleshooting, and preventive maintenance.

Operating a WRRF is challenging and requires continuing education to keep up with those challenges. As such, this study guide contains challenging questions and detailed solutions. These questions can be used to help develop advanced knowledge and ensure that WRRFs are fulfilling their mission of environmental protection. A list of symbols and acronyms, conversion factors, and a glossary are also included in this study guide.

The following authors are responsible for revising this edition of the study guide:

Bulbul Ahmed, Ph.D., P.E.	(4, 9, 17)
Jeanette A. Brown, P.E., BCEE	(10, 27, 31, 32)
Viraj deSilva, Ph.D., P.E., BCEE	(19, 22, 26)
Georgine Grissop, P.E., BCEE	(5, 14, 15, 16)
Natasha Jensen, EIT	(6, 7, 24)
Murthy Kasi, Ph.D., P.E.	(20, 28, 30, 33)
Isaiah LaRue	(8, 12, 18, 21)
David M. Mason	(3, 23, 25, 29)
Nashita Naureen	(2, 11, 13)

Authors' and reviewers' efforts were supported by the following organizations:

Black & Veatch, Indianapolis, Indiana
Carollo Engineers, Inc., Phoenix, Arizona
CDM Smith Inc., Boston, Massachusetts
City of Franklin, Tennessee
DC Water, Washington, District of Columbia
Dillon Consulting Limited, London, Ontario, Canada
GE Power & Water, The Woodlands, Texas
HDR Inc.
Manhattan College, Riverdale, New York

Smith & Loveless, Lenexa, Kansas
Structure Point, Indianapolis, Indiana
Toronto Water, Toronto, Ontario, Canada
Tulane University, New Orleans, Louisiana
University of Wisconsin, Madison
Xylem, Inc., Brown Deer, Wisconsin

Chapter 2
PERMIT COMPLIANCE AND WASTEWATER TREATMENT SYSTEMS

Problems

PROBLEM 2.1

What are the four guiding principles of the Clean Water Act?

PROBLEM 2.2

National Pollutant Discharge Elimination System (NPDES) permits typically specify the discharge location, allowable discharge flows, allowable concentrations or mass loads of pollutants in the discharge, limits of the mixing zone (if any), and monitoring and reporting requirements.

    a)   True
    b)   False

PROBLEM 2.3

National Pollutant Discharge Elimination System permits for secondary treatment facilities typically have average monthly effluent limits for five-day biochemical oxygen demand (BOD_5) and total suspended solids (TSS) of

    a)   10 and 30 mg/L, respectively.
    b)   20 and 30 mg/L, respectively.
    c)   30 and 30 mg/L, respectively.

PROBLEM 2.4

What is regulated by 40 CFR Part 503?

a) Stabilization
b) Sludge used for landfilling
c) Biosolids used for land application

PROBLEM 2.5

The list of 126 priority pollutants can be divided into four groups. They are

a) Volatile organic compounds, volatile inorganic compounds, hydrocarbons, and polychlorinated pesticides.
b) Heavy metals and cyanide, volatile organic compounds, semivolatile organic compounds, and pesticides and polychlorinated biphenyls.
c) Heavy metals and cyanide, volatile inorganic compounds, polycarbonated hydrocarbons, and semivolatile inorganic compounds.

PROBLEM 2.6

The primary objective of a water resource recovery facility (WRRF) is to process wastewater to meet permit limits and to not damage the environment.

a) True
b) False

PROBLEM 2.7

Combined sewer overflows suggest that the collection system is a separate system.

a) True
b) False

PROBLEM 2.8

Pumping stations are only necessary in large cities.

a) True
b) False

PROBLEM 2.9

Name six considerations that regulators take into account when developing a permit for a particular city.

PROBLEM 2.10

The Occupational Safety and Health Administration was created to protect

a) Public health and welfare.
b) Worker safety and health.
c) Public safety and health.

PROBLEM 2.11

Name two processes that remove water from sludge.

a) Conditioning and thickening
b) Dewatering and flocculation
c) Dewatering and thickening

PROBLEM 2.12

The State Revolving Fund is important to WRRFs (publicly owned treatment works is used in the regulations) because it provides

a) Low-interest loans to towns to build and upgrade WRRFs.
b) Construction grants for 100% of the cost of building and upgrading WRRFs.
c) Safe debt limits for WRRFs to borrow money.

PROBLEM 2.13

Draw a box diagram showing the typical flow through a secondary treatment facility.

PROBLEM 2.14

Stabilization may result in a material that can be safely used on land as a soil conditioner.

a) True
b) False

PROBLEM 2.15

A WRRF operating under a State Pollutant Discharge Elimination System permit

a) Is not required to meet secondary standards.

b) Will have the same limits or more stringent limits as a facility operating under an NPDES permit.

c) Is in a state that does not have an approved regulatory program.

PROBLEM 2.16

An environmental management system is designed to

a) Integrate environmental issues to an organization management plan.

b) Integrate emergency medical systems to respond to facility injuries.

c) Integrate environmental management to a financial plan.

PROBLEM 2.17

Liquid treatment processes include thickening and dewatering.

a) True

b) False

PROBLEM 2.18

Advanced treatment implies that

a) The facility is designed to remove organic compounds.

b) The facility is designed to remove nutrients, trace organics, and/or color.

c) The facility is designed to remove high levels of coliform.

PROBLEM 2.19

Primary treatment removes _____% BOD and _____% TSS.

a) 60 and 30, respectively

b) 60 and 90, respectively

c) 30 and 60, respectively

PROBLEM 2.20

Belt filter presses are used for what purpose?

Solutions

SOLUTION 2.1

(1) No one has the right to pollute U.S. waters, so a permit is required to discharge any pollutant;

(2) Permits limit the types and concentrations of pollutants allowed to be discharged, and permit violations can be punished by fines and imprisonment;

(3) Some industrial permits require companies to use the best treatment technology available, regardless of the receiving water's assimilative capacity; and

(4) Pollutant limits involving more waste treatment than technology-based levels, secondary treatment (for municipalities), or best practicable technology (for industries) are based on waterbody-specific water quality standards.

SOLUTION 2.2

a) True

SOLUTION 2.3

c) 30 and 30 mg/L, respectively

SOLUTION 2.4

c) Biosolids used for land application

SOLUTION 2.5

b) Heavy metals and cyanide, volatile organic compounds, semivolatile organic compounds, and pesticides and polychlorinated biphenyls.

SOLUTION 2.6

a) True

SOLUTION 2.7

b) False

SOLUTION 2.8

b) False

SOLUTION 2.9

Any six of the following:

- Preventing disease,
- Preventing nuisances,
- Protecting drinking water supplies,
- Conserving water,
- Maintaining navigable waters,
- Protecting waters for swimming and recreational use,
- Maintaining healthy habitats for fish and other aquatic life, and
- Preserving pristine waters to protect ecosystems.

SOLUTION 2.10

b) Worker safety and health.

SOLUTION 2.11

c) Dewatering and thickening

SOLUTION 2.12

a) Low-interest loans to towns to build and upgrade WRRFs.

SOLUTION 2.13

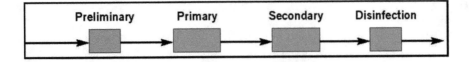

SOLUTION 2.14

a)   True

SOLUTION 2.15

b)   Will have the same limits or more stringent limits as a facility operating under an NPDES permit.

SOLUTION 2.16

a)   Integrate environmental issues to an organization management plan.

SOLUTION 2.17

b)   False

SOLUTION 2.18

b)   The facility is designed to remove nutrients, trace organics, and/or color.

SOLUTION 2.19

c)   30 and 60, respectively

SOLUTION 2.20

Dewatering

Chapter 3
FUNDAMENTALS OF MANAGEMENT

Problems

PROBLEM 3.1

The fundamental mission of a water resource recovery facility (WRRF) is to protect the health of the public and the environment.

    a) True
    b) False

PROBLEM 3.2

Which of the following are typical asset management strategies?

    a) Know the condition and criticality of all assets
    b) Know the remaining useful life of an asset
    c) Know the performance and failure history of an asset
    d) Know the risks and consequences of failure
    e) All of the above

PROBLEM 3.3

Which of the following is not a significant pathway to an integrated strategic management approach?

    a) Outstanding leadership
    b) Thoughtful planning
    c) Having a good relationship with the media
    d) Long-term operating strategy
    e) Business-like management of assets
    f) Stakeholder support and alignment

PROBLEM 3.4

Which of the following is not an essential reason why an operations and maintenance (O&M) plan is important?

a)   It is the foundation for an effective operations strategy.

b)   It serves as the basis for managing a facility's operation.

c)   It is used to organize, administer, and respond to daily operations elements at the facility.

d)   It provides a defense for salary and/or benefit increases.

e)   It defines procedures for handling day-to-day activities and emergency response.

PROBLEM 3.5

The staffing section of your facility's O&M plan includes the following:

a)   A description of the structure of the organization

b)   What reporting relationships exist

c)   Procedures for handling normal operations and emergency and unusual situations

d)   All of the above

PROBLEM 3.6

Which of the following are common elements to consider that can support effective personnel management?

a)   Recruitment

b)   Training and development

c)   Retention and succession planning

d)   Meaningful performance appraisal and discipline

e)   All of the above

PROBLEM 3.7

The manager should leverage and build relationships with external stakeholders, including the following:

a)   Facility neighbors

b)   Police, fire, and other city or utility departments

c)   Regulatory agencies and professional organizations

d) Media (an external media communications policy and strategy should be developed before engaging with the media)

e) All of the above

PROBLEM 3.8

Achieving efficiency improvements requires constant attention to discrete practices, functions, subgroups, and technologies within the organization.

a) True
b) False

PROBLEM 3.9

Process benchmarking focuses on improving internal programs and processes by learning how the "best" organizations carry them out.

a) True
b) False

PROBLEM 3.10

Sound fiscal management includes which of the following elements?

a) Service standards and performance levels
b) Informed asset management decisions
c) Understanding the financial implications of budgeting decisions (capital and operating costs)
d) Monitoring and managing the budget (actual costs vs budgeted costs by month and by quarter)
e) All of the above

PROBLEM 3.11

Regulatory permits required by most wastewater utilities include which of the following?

a) Wastewater discharge
b) Air emissions
c) Building modifications
d) Underground tanks
e) All of the above

PROBLEM 3.12

Preventive and predictive maintenance involve all but which of the following?

a)   Performing maintenance in advance of failure

b)   Performing maintenance on an asset that is failing or has failed

c)   Monitoring the condition of equipment and selecting the most appropriate time to service it

d)   Ensuring that inventory is available when and where needed so that labor is not wasted waiting for, or looking for, parts

PROBLEM 3.13

An organization chart is useful for which of the following?

a)   Describe the structure of the organization

b)   Describe the reporting relationships

c)   Describe who is responsible for what task

d)   All of the above

PROBLEM 3.14

Managers should avoid diagnosing personal problems for employees that have external issues that affect their work performance.

a)   True

b)   False

PROBLEM 3.15

The facility manager should follow a disciplinary process that includes all of the following except

a)   Pertinent facts are collected in advance of a disciplinary discussion and presented fairly.

b)   Discussions taking place with the union steward present.

c)   Posting names and alleged infractions on the union bulletin board.

d)   Notices are given and hearings are scheduled in a timely fashion with adequate time to respond.

e)   A clear understanding of which rule(s)/regulation(s)/contract provision(s) are allegedly violated.

PROBLEM 3.16

The manager may generally stimulate motivation by creating a work environment that includes the following elements:

a) Make responsibilities known and reasonable
b) Make sure that the reasons for work are clear to those who do the work
c) Find rewards for good performance
d) All of the above

PROBLEM 3.17

Methods used to develop the teamwork that is essential for a successful facility include the following:

a) Arranging for group effort to solve a problem, to develop a budget for the next year, or to undertake another project
b) Having employees share their knowledge and goals with one another
c) Providing group and individual awards based on group success and individual contributions in the group
d) Rotating jobs or sharing jobs so employees gain a better understanding of other situations
e) All of the above

PROBLEM 3.18

The emergency planning process includes which of the following?

a) Identifies hazards and needs
b) Sets goals, determines objectives, and sets priorities
c) Designs action programs
d) Provides plan implementation strategies and evaluates results
e) All of the above

PROBLEM 3.19

Elements of sound fiscal management include which of the following?

a) Define service standards and performance levels.
b) Make informed decisions on asset management.
c) Understand the financial implications of budgeting decisions.

    d)   Ensure that the budget is monitored and managed.

    e)   All of the above

PROBLEM 3.20

Data management technologies at the facility level include which of the following?

    a)   Spreadsheets

    b)   Laboratory information management systems (LIMS)

    c)   Computerized maintenance management systems (CMMS)

    d)   All of the above

Solutions

SOLUTION 3.1

    a)   True

The fundamental mission of a WRRF is to protect the health of the public and the environment.

The significant responsibility of ensuring that the facility is in accordance with specified regulatory and other local requirements lies squarely on the manager's shoulders.

SOLUTION 3.2

    e)   All of the above

SOLUTION 3.3

    c)   Having a good relationship with the media

The main pathways to the integrated strategic management approach are to develop an outstanding leadership team, engage in thoughtful planning, develop a long-term operating strategy, manage assets from a business-like perspective, and engage mechanisms that develop stakeholder support and alignment.

SOLUTION 3.4

    d)   It provides a defense for salary and/or benefit increases.

The O&M plan is considered the core management tool for the facility manager to organize, administer, and respond to daily operational elements at the facility.

The O&M plan not only defines the roles and responsibilities of staff, but also establishes procedures for handling everyday situations and emergencies.

By formalizing standard operating procedures, the O&M plan also provides a framework that facilitates periodic reviews and updates in response to changes.

SOLUTION 3.5

d)   All of the above

The staffing section of an O&M plan should describe the structure of the organization, what reporting relationships exist, and who is responsible for what task.

SOLUTION 3.6

e)   All of the above

The following are common elements to consider that can support effective personnel management: personnel management system, recruitment, training and development, retention and succession planning, meaningful performance appraisal, and discipline.

SOLUTION 3.7

e)   All of the above

The manager should leverage and build relationships with external stakeholders, including the following:

- Facility neighbors;
- Vendors;
- Police department;
- Fire department (especially for emergency response planning and preparation);
- Regulatory agencies;
- Other city or utility departments (e.g., streets and roads, parks, and legal);
- Media (an external media communications policy and strategy should be developed before engaging with the media); and
- Professional organizations (e.g., community groups and local and state Water Environment Federation affiliates).

SOLUTION 3.8

a) True

Achieving efficiency improvements requires constant attention to discrete practices, functions, subgroups, and technologies within the organization.

SOLUTION 3.9

a) True

Process benchmarking is a more effective way to apply benchmarking efforts in the public arena because it focuses on improving internal programs and processes by learning how "the best" organizations conduct such activities. In process benchmarking, utility practices and other "non-numerical" items such as procedures and systems are compared to others.

SOLUTION 3.10

e) All of the above

SOLUTION 3.11

e) All of the above

Regulatory agencies and the public expect compliance with permits that may include wastewater discharges, air emissions, building modifications, underground tanks, and other items.

SOLUTION 3.12

b) Performing maintenance on an asset that is failing or has failed

Maintenance of equipment that has failed or is nearing failure constitutes corrective maintenance.

SOLUTION 3.13

d) All of the above

The staffing section of an O&M plan should describe the structure of the organization, what reporting relationships exist, and who is responsible for what task. An organization chart should be used to convey this information.

SOLUTION 3.14

a)   True

Managers should avoid diagnosing personal problems for employees that have external issues that affect their work performance. However, they should point out to the employee that there are sources available to them for assistance, such as an employee assistance program. A program that includes professional counseling is effective at helping employees overcome difficult personal problems and becoming motivated to work.

SOLUTION 3.15

c)   Posting names and alleged infractions on the union bulletin board

The facility manager should follow a disciplinary process that includes the following:

- Train all supervisors in disciplinary procedures;
- Ensure that discipline is administered evenly and respectfully;
- Require that investigations, notices, and hearings occur in a timely fashion;
- Ensure that employees understand performance requirements and disciplinary steps; and
- Document performance and discipline and file the records.

SOLUTION 3.16

d)   All of the above

In addition, the manager may generally stimulate motivation by creating a work environment that includes the following elements:

- Make responsibilities known and reasonable. Match authority to responsibilities. Eliminate the frustrations of slow approvals, long meetings, and lack of follow-up;
- Find rewards for good performance. Often, prompt recognition by the boss is the easiest, most effective reward;
- The reasons for work must be clear to those who do the work. With shift work, this requires supervisors to visit all shifts and leave instructions that include reasons for the work;
- Working relationships must be comfortable and encouraging for teamwork. Social events, time out for some fun event at work, and fair and equal treatment can help build an environment that encourages good performance; and
- Recognition that continued education and training are essential in the modern work environment, which is characterized by exponential growth in technical knowledge required to improve productivity with decreased resources.

SOLUTION 3.17

e) All of the above

Methods used to develop the teamwork that is essential for a successful facility include the following:

- Arranging for group effort to solve a problem, to develop a budget for the next year, or to undertake another project. As a requirement for such a team effort, everyone must be able to contribute and must have a stake in the outcome;
- Having employees share their knowledge and goals with one another;
- Providing group and individual awards based on group success and individual contributions in the group;
- Rotating jobs or sharing jobs so employees gain a better understanding of other situations; and
- Documenting experiences and lessons learned for future reference and knowledge retention.

Instilling an atmosphere that is rewarding and enjoyable to work in.

SOLUTION 3.18

e) All of the above

The planning process identifies hazards and needs, sets goals, determines objectives, sets priorities, designs action programs, provides plan implementation strategies, evaluates results, and repeats the steps.

SOLUTION 3.19

a) All of the above

SOLUTION 3.20

d) All of the above

At the facility level, data management technologies can include spreadsheets, an LIMS, and a CMMS. Of these technologies, organizations seem to struggle most with fully implementing and using the capabilities of a CMMS.

Chapter 4
PRETREATMENT PROGRAM REQUIREMENTS FOR INDUSTRIAL WASTEWATER

Problems

PROBLEM 4.1

The Clean Water Act was established in which year?

a) 1972
b) 1948
c) 1956

PROBLEM 4.2

The General Pretreatment Regulations apply to all non-domestic sources that introduce pollutants to a water resource recovery facility (WRRF; publicly owned treatment works is used in the regulations).

a) True
b) False

PROBLEM 4.3

List three types of discharge standards needed to meet the objectives of the National Pretreatment Program.

PROBLEM 4.4

Prohibited discharge standards are enforced and applicable at which level?

a) National
b) State
c) Local

PROBLEM 4.5

The compliance for categorical pretreatment standards is measured at the end of pipe discharge point for the industrial facility.

a) True
b) False

PROBLEM 4.6

The compliance for local limits is typically measured at the end of pipe discharge point for the industrial facility.

a) True
b) False

PROBLEM 4.7

Generally, which of the following acts as the pretreatment "control authority" with respect to industrial users that discharge to a WRRF?

a) State
b) U.S. Environmental Protection Agency (U.S. EPA)
c) WRRF

PROBLEM 4.8

Who is responsible for ensuring that local program implementation is consistent with all applicable federal requirements and is effective in achieving the National Pretreatment Program's goals?

PROBLEM 4.9

In general, which type of sampling is recommended for collecting samples from an industrial facility?

a) Grab sampling
b) Flow-proportional composite sampling
c) Time-based composite sampling

PROBLEM 4.10

It is the responsibility of which of the following authorities to publish a list of industries that were in significant non-compliance with applicable pretreatment standards in the past 12 months?

a) U.S. EPA
b) Federal government
c) Local authorities

PROBLEM 4.11

Submitting the baseline pretreatment monitoring report is the responsibility of which of the following?

a) WRRF
b) Industrial user
c) State

PROBLEM 4.12

Which of the following processes can remove targeted constituents from industrial wastestreams without producing any change in the structure of the constituent?

a) Chemical processes
b) Physical processes
c) Biological processes

PROBLEM 4.13

Which of the following treatment processes can remove organic compounds from industrial wastestreams?

a) Stripping by air
b) Adsorption on activated carbon
c) Advanced oxidation processes
d) All of the above

PROBLEM 4.14

State at least two criteria used to classify an industrial user as a significant industrial user.

PROBLEM 4.15

The maximum allowable headworks loading method is one of the commonly used methods to derive which of the following?

a) Prohibited discharge standard
b) Categorical pretreatment standard
c) Local limits

Solutions

SOLUTION 4.1

a) 1972

SOLUTION 4.2

a) True

SOLUTION 4.3

- Prohibited discharge,
- Categorical pretreatment, and
- Local limits.

SOLUTION 4.4

a) National

SOLUTION 4.5

b) False

SOLUTION 4.6

a) True

SOLUTION 4.7

c) WRRF

SOLUTION 4.8

The approval authority

SOLUTION 4.9

b) Flow-proportional composite sampling

SOLUTION 4.10

c) Local authorities

SOLUTION 4.11

b) Industrial user

SOLUTION 4.12

b) Physical processes

SOLUTION 4.13

d) All of the above

SOLUTION 4.14

(1) An industrial user discharges an average of 95 m³/d (25 000 gpd) or more of process wastewater to the WRRF and

(2) An industrial user contributes a process wastestream making up 5% or more of the average dry weather hydraulic or organic capacity of the WRRF.

SOLUTION 4.15

c) Local limits

Chapter 5
SAFETY

Problems

PROBLEM 5.1

Which of the following hazards is not typically found in water resource recovery facilities?

a) Electrical
b) Bacteriological
c) Combustible grain dust

PROBLEM 5.2

Workers are sometimes careless because

a) They think, "It is not their problem".
b) They may think, "Nobody will know if I do ...".
c) They are distracted by personal problems when working.
d) All of the above

PROBLEM 5.3

Which of the following hazardous chemicals is typically found in water resource recovery facilities?

a) Sodium pentothal
b) Sodium hydroxide
c) Ferric nitrate

PROBLEM 5.4

Hydrogen sulfide can cause death in a few minutes at a concentration of 0.2%.

a) True
b) False

PROBLEM 5.5

Whenever details of hazardous chemical storage become part of the public record, plans for security of these chemicals do not need to be made.

a) True
b) False

PROBLEM 5.6

To safely handle chemicals, what is it important to have?

a) A truck
b) A supervisor
c) Personal protective equipment

PROBLEM 5.7

In the event of a chlorine leak, where is it recommended that a self-contained breathing apparatus be located?

a) Outside the chlorine room
b) In the operations office
c) In your truck

PROBLEM 5.8

To protect themselves from aerosols, wastewater workers could wear a particle mask to protect them from what?

a) Many airborne contaminants
b) Identification theft
c) Slips, trips, and falls

PROBLEM 5.9

An easily overlooked, but highly effective way to protect your health is to what?

PROBLEM 5.10

During a confined space entry, the duty of the attendant during an emergency is to what?

PROBLEM 5.11

Name one of the three conditions that characterize an accessible area as a confined space.

PROBLEM 5.12

According to Occupational Safety and Health Administration (OSHA) regulations, it is permissible for small communities to pool their resources to perform work in confined spaces.

    a)   True
    b)   False

PROBLEM 5.13

While trenching and excavating, 0.76 m³ (27 cu ft) of soil can weigh more than 1400 kg (1.5 tons).

    a)   True
    b)   False

PROBLEM 5.14

All previously disturbed soil is automatically classified as either Class B or Class C soil. If an excavation occurred to either replace or repair pipe, it is typically classified as which?

    a)   Class A
    b)   Class B
    c)   Class C

PROBLEM 5.15

In facilities with laboratories, it is a basic safety practice to provide an emergency eyewash station and shower.

    a) True
    b) False

PROBLEM 5.16

The most popular description of combustion is the fire triangle. What are the three sides of the triangle?

PROBLEM 5.17

When using a fire extinguisher, what does the memory device of PASS stand for?

PROBLEM 5.18

Hydraulic systems may store energy even when deenergized. This form of energy is known as what?

    a) Latent energy
    b) Thermodynamics
    c) Hydroelectric energy

PROBLEM 5.19

When performing collection system maintenance and repairs, workers should be protected with a traffic control zone.

    a) True
    b) False

PROBLEM 5.20

Responsibility for providing training and a written safety and health policy belongs to whom?

    a) City council
    b) Management
    c) Equipment suppliers

PROBLEM 5.21

It is acceptable to use a tool incorrectly if it saves time.

a) True

b) False

PROBLEM 5.22

Training serves as a preventive measure against accidents and job-related illness. It starts when a new employee is hired and never ends. Safety is a joint effort between management and labor and requires commitment from both.

a) True

b) False

Solutions

SOLUTION 5.1

c)   Combustible grain dust

SOLUTION 5.2

d)   All of the above

SOLUTION 5.3

b)   Sodium hydroxide

SOLUTION 5.4

a)   True

SOLUTION 5.5

b)   False

SOLUTION 5.6

c)   Personal protective equipment

SOLUTION 5.7

a)   Outside the chlorine room

SOLUTION 5.8

a)   Many airborne contaminants

SOLUTION 5.9

Wash your hands

SOLUTION 5.10

Summon rescue

SOLUTION 5.11

A confined space is characterized by the following three conditions:

(1)   Limited entrance and egress
(2)   Unfavorable neutral ventilation (stale air)
(3)   A design that allows for limited occupancy

SOLUTION 5.12

a)   True

SOLUTION 5.13

a)   True

SOLUTION 5.14

c)   Class C

SOLUTION 5.15

a)   True

SOLUTION 5.16

Oxygen, fuel, and heat

SOLUTION 5.17

Pull, aim, squeeze, sweep

SOLUTION 5.18

a)   Latent energy

SOLUTION 5.19

a)   True

SOLUTION 5.20

b)   Management

SOLUTION 5.21

b)   False

SOLUTION 5.22

a)   True

Chapter 6
MANAGEMENT INFORMATION SYSTEMS— REPORTS AND RECORDS

Problems

PROBLEM 6.1

With a properly designed computer system, data

a) Are entered several times.
b) Are difficult to analyze.
c) Are inconsistent across different computer applications.
d) Are entered one time and used for multiple purposes.

PROBLEM 6.2

Management information systems can be separated into which of the following main categories:

a) Hardware, software, data, and vendor applications
b) Hardware, software, data, and business processes
c) Reports, hardware, and service level agreements

PROBLEM 6.3

List three software applications typically used by wastewater agencies.

PROBLEM 6.4

Describe three information technology management issues that utilities may experience when adopting new software applications.

PROBLEM 6.5

Define *information technology governance*.

PROBLEM 6.6

List three benefits of effective information technology governance.

PROBLEM 6.7

Identify three aspects of an information technology plan.

PROBLEM 6.8

Converting data from an existing system to a new system typically involves "cleansing" the data.

    a)   True
    b)   False

PROBLEM 6.9

Name three components of a typical computerized maintenance management system (CMMS).

PROBLEM 6.10

Describe an annual operating report.

PROBLEM 6.11

The software provider is responsible for any programming problems that might produce errors in regulatory reports.

    a)   True
    b)   False

PROBLEM 6.12

List three required features of an effective record management system.

PROBLEM 6.13

Identify three uses for intranet-based applications for utilities.

Solutions

SOLUTION 6.1

d)   Are entered one time and used for multiple purposes.

SOLUTION 6.2

b)   Hardware, software, data, and business processes

SOLUTION 6.3

Any three of the following software applications are typically used by wastewater agencies:

- Computerized maintenance management systems (may also be called work management systems or enterprise asset management systems);
- Geographic information systems;
- Laboratory information management systems;
- Human resources management systems;
- Financial information systems;
- Process control systems;
- Supervisory control and data acquisition systems;
- Modeling software (e.g., a model of the collection system); and
- Personal productivity systems.

SOLUTION 6.4

Three information technology management issues that utilities may experience when adopting new software applications include any of the following:

- Integrating enterprise applications,
- Aligning business processes with technology,

- Ensuring that information technology addresses business requirements,
- Providing adequate training for end-users and support staff,
- Providing sufficient support resources, and
- Standardization.

SOLUTION 6.5

Information technology governance is a method by which technology decisions and investments are planned and progress is tracked to ensure that the client's needs are understood and addressed.

SOLUTION 6.6

Three benefits of effective information technology governance include any the following:

- Information technology strategies are aligned with business strategies;
- An effective structure (technological and organizational) is in place to facilitate implementing the strategy;
- Goals accurately reflect the needs of the client and are articulated clearly and defined;
- Information technology issues are understood;
- Risk is managed;
- Goals are measured and status is reported; and
- The final results are audited, allowing for lessons learned to become part of the institutional knowledge and future planning to be informed.

SOLUTION 6.7

Three aspects of an information technology plan include any of the following:

- Defines the agency's information technology direction and standards,
- Lists significant information technology projects (including schedule and costs),
- Establishes important standards,
- Addresses required utility support for information technology (such as training),
- Requires a cross-functional perspective, and
- Requires updating every year.

SOLUTION 6.8

a) True

SOLUTION 6.9

Three components of a typical CMMS include any of the following:

- List of assets (equipment inventory);
- Lists of activities and procedures;
- Purchasing/procurement information;
- Inventory/storeroom/warehouse information including forcasting;
- Work orders (corrective, preventive, and predictive maintenance);
- Work schedules and wages;
- Skills tracking; and
- Reports and analyses.

SOLUTION 6.10

An annual report summarizes the developments and activities of the agency for the preceding year. The report will include sections with general information, operations, maintenance, and improvements carried out including a description of the capital improvement program. A description of the treatment process and how it works should be included, as well as current, past, and projected future hydraulic capacity and loadings. Operating efficiency and comments on past and present water quality are also important aspects. The report will also contain financial information, such as the facility's financial status, operating costs, and capital project costs.

SOLUTION 6.11

b)   False

The permit holder is ultimately responsible for the accuracy of regulatory reports.

SOLUTION 6.12

Three required features of an effective record management system include any of the following:

- Convenient accessibility,
- Secure storage of data,
- Regular system and data backups, and
- Training on use of the applications.

SOLUTION 6.13

Three uses for intranet-based applications for utilities include any of the following:

- Sharing information regarding upcoming events,
- Access to utility applications such as work management systems or geographic information systems,
- A single reference location for data of interest to all employees, and
- Access to electronic versions of physical facility records.

Chapter 7
PROCESS INSTRUMENTATION

Problems

PROBLEM 7.1

A specification for an online chlorine analyzer requires that the analyzer have a span of 0 to 10 mg/L, a minimum detection limit of 0.05 mg/L, an accuracy of 3% of span, and a repeatability of 61.0% of value. When this analyzer is installed in chlorinated facility effluent and the true chlorine concentration reported by the analyzer is 3.8 mg/L, the analyzer output will be what if it is performing to specification?

    a)   3.7 to 3.9 mg/L
    b)   3.75 to 3.85 mg/L
    c)   3.8 mg/L
    d)   3.5 to 4.1 mg/L
    e)   3.686 to 3.914 mg/L

PROBLEM 7.2

Which of the following is not an analytical sensor?

    a)   An amperometric chlorine analyzer
    b)   A colorimetric nitrate analyzer
    c)   A galvanic dissolved oxygen sensor
    d)   An ultrasonic level sensor
    e)   A nephelometric turbidity sensor

PROBLEM 7.3

A water resource recovery facility is selecting a flow meter to measure the flowrate of treated effluent. The flow measurement will be made in an open channel that transports the treated effluent to the outfall. A number of suppliers have provided the purchasing department with quotes on flow meters for this application. Which of the following meters should not be selected for this application?

a) A magnetic flow meter

b) A V-notch weir and ultrasonic-level sensor

c) A Parshall flume and bubbler tube

d) A rectangular weir with a capacitance probe

PROBLEM 7.4

An online dissolved oxygen analyzer has a span of 0 to 20 mg/L and the transmitter provides an output signal of 4 to 20 mA. The output from the transmitter is 8 mA. The temperature of the water being measured is 15 °C. What is the dissolved oxygen concentration in the water if the instrument has been calibrated at 10 °C so that the output is 4 mA when the dissolved oxygen is 0 mg/L and the output is 20 mA when the dissolved oxygen is 10 mg/L?

PROBLEM 7.5

A utility wishes to reduce its energy costs by reducing the speed of the positive-displacement blowers that are used to provide air to the process at times when they are providing air volumes in excess of what is required by the microorganisms. Which of the following would be an appropriate instrumentation and process control strategy?

a) Measure the mixed liquor suspended solids (MLSS) concentration with an ultrasonic solids sensor and increase the speed of the blowers in proportion to the MLSS concentration.

b) Measure the phosphorus concentration in the secondary clarifier effluent using a colorimetric analyzer and decrease the speed of the blowers when the phosphorus concentration exceeds a preset concentration.

c) Measure the sludge blanket level in the secondary clarifiers and increase the speed of the blowers when the blanket level increases.

d) Measure the airflow rate provided by the blowers with an orifice plate and pressure transducers and reduce the speed of the blowers when the airflow rate exceeds a preset rate.

e) Measure the dissolved oxygen concentration in the bioreactor and reduce the speed of the blowers when the dissolved oxygen concentration is above a preset concentration.

PROBLEM 7.6

A water resource recovery facility must remove residual chlorine from the effluent before discharge to eliminate toxicity associated with chlorine. The facility has installed an online chlorine sensor to measure the concentration of chlorine remaining in the wastewater after sulfur dioxide has been added to remove the chlorine residual. The output from the chlorine residual analyzer is used to control the rate of addition of the sulfur dioxide. This is an example of what type of control?

a) Manual control
b) Feedforward control
c) Feedback control
d) Compound loop control
e) Advanced control

PROBLEM 7.7

Which of the following is not a primary element or a sensor in an instrumentation and control system?

a) An ultrasonic device measuring the level in a biosolids storage tank
b) An ion-select electrode measuring ammonia concentration in facility effluent
c) An open/closed position indicator on a valve
d) A variable-frequency drive that changes the speed of a positive-displacement blower drive
e) A tachometer generator measuring the speed of a positive-displacement blower drive

PROBLEM 7.8

A magnetic flow meter is calibrated for a span of 100 to 500 L/s (1600 to 8000 gpm) and the local transmitter/indicator indicates the flow as percent of span. When the operator records the flow in the log book, the local indicator shows a flow reading of 35% of span. What is the flow in liters per second (L/s) (gallons per minute [gpm]) at the time that the operator took the reading?

PROBLEM 7.9

Which of the following would not be part of an automated control system to control the dissolved oxygen concentration in a bioreactor?

a) A dissolved oxygen sensor
b) A 4- to 20-mA signal transmitter
c) An airflow meter measuring the airflow to the bioreactor tank
d) An airflow control valve
e) A suspended solids sensor measuring the concentration of MLSS in the bioreactor

PROBLEM 7.10

A water resource recovery facility uses a rectangular weir to measure the treated effluent flow. An ultrasonic sensor is used for measurement of liquid levels to determine the flow over the weir. The expected maximum flow to the water resource recovery facility is 40 000 m³/d (10.6 mgd). At this flow, the maximum water depth in the channel upstream of the weir is 1.5 m (4.9 ft). The weir crest is at an elevation of 1.0 m (3.3 ft) above the channel floor. How far (in meters) upstream of the weir should the ultrasonic sensor be positioned?

PROBLEM 7.11

The measurement of pH in a water resource recovery facility is critical to maintain an environment for microorganisms to survive. Instrumentation is important in the accurate measurement of this type. What type of measurement is pH considered?

a) A type of substance
b) A group of chemicals
c) An effect
d) A specific chemical ion
e) A compound substance

PROBLEM 7.12

At a lift station, three variable-speed pumps are used to send wastewater to the water resource recovery facility. The facility is located only one block away from the lift station. The lift station serves several 24-hour businesses so the incoming flow to the station is constant and consistent. Assuming all necessary process measurement could be provided, what control method would be best suited to provide stable, straightforward control of all three pumps?

a) On–off control
b) Reset control action
c) Proportional control
d) Three-mode control
e) Simple control

PROBLEM 7.13

All process measurements are extremely important and can easily justify the purchase of online instrumentation.

a) True
b) False

PROBLEM 7.14

With the advent of programmable logic controllers and other sophisticated control devices, there is no more need to inspect equipment.

a) True
b) False

PROBLEM 7.15

Fill in the blank: Instrument error is the difference between the instrument reading and the _____ value.

PROBLEM 7.16

Fill in the blank: Differential head flow meters use an in-pipe constriction that produces a temporary and measurable _____ across it.

PROBLEM 7.17

A nephelometer is a meter that

a) Measures the amount of scattered light through a sample.
b) Infers the size and concentration of suspended particulates.
c) Is a photoelectric device.
d) All of the above
e) None of the above

PROBLEM 7.18

A water resource recovery facility needs to purchase a turbidimeter for a certain new effluent stream. The anticipated turbidity of the effluent will have a maximum of 100 NTU and a minimum of 80 NTU. A new filter treatment process will be added in the future that will improve the turbidity of the effluent and result in an anticipated maximum reading of 10 NTU and a minimum of 5 NTU. Accurate readings of turbidity will be extremely important in this process. Once the new filter is in place, should a new turbidimeter be purchased, or should the facility staff keep the existing one to save money?

PROBLEM 7.19

Fill in the blank: A reasonability check is the act of observing the indication of an instrument with process _____ and compared with other related instrumentation.

Solutions

SOLUTION 7.1

d) 3.5 to 4.1 mg/L

If the analyzer meets specification, it will measure the true value of chlorine concentration to within 3% of span. The instrument span is 0 to 10 mg/L. Therefore, the measurement should be accurate to 3% × (10 mg/L − 0 mg/L) = 0.3 mg/L.

If the true chlorine concentration is 3.8 mg/L, then the chlorine analyzer meets the specification if the output is

$$3.8 \text{ mg/L} \pm 0.3 \text{ mg/L, or between 3.5 and 4.1 mg/L.}$$

The minimum detection limit and repeatability are both important considerations in selecting an analyzer, but neither factor into the determination of what analyzer inputs fall within accuracy specifications.

SOLUTION 7.2

d) An ultrasonic level sensor

Level indicators are physical sensors.

SOLUTION 7.3

a) A magnetic flow meter

Magnetic flow meters require a full pipe for proper operation. These flow meters measure the fluid velocity, and assume a full pipe to convert the velocity to a volumetric flow.

SOLUTION 7.4

2.5 mg/L dissolved oxygen

The instrument is calibrated so that the output is 4 mA when the dissolved oxygen is 0 mg/L and 20 mA when the dissolved oxygen is 10 mg/L. Therefore, the change in the output represents a change in dissolved oxygen of

$$(10 \text{ mg/L} - 0 \text{ mg/L}) / (20 \text{ mA} - 4 \text{ mA}) = 0.625 \text{ mg/L per mA.}$$

If the output of the transmitter is 8 mA, the dissolved oxygen is

$$(8 \text{ mA} - 4 \text{ mA}) \times 0.625 \text{ mg/L per mA} = 2.5 \text{ mg/L dissolved oxygen.}$$

Although temperature does affect the capacity of oxygen to dissolve in the liquid, it does not matter at which temperature the analyzer was calibrated. Most analyzers have built-in compensation for temperature.

SOLUTION 7.5

e) Measure the dissolved oxygen concentration in the bioreactor and reduce the speed of the blowers when the dissolved oxygen concentration is above a preset concentration.

This was the only option that measured the residual or excess oxygen (feedback control). It will best account for all existing possible errors and process disturbances (i.e., a cumulative effect), whereas the other options only explore the effect of one disturbance and depend on a model of how that disturbance could affect the process.

SOLUTION 7.6

c) Feedback control

In this case, the error (actual chlorine residual vs a setpoint) is used to affect the control variable (addition rate of sulfur dioxide). Possible disturbances could include a change in effluent flowrate or a change in chlorine dosing/demand.

SOLUTION 7.7

d) A variable-frequency drive that changes the speed of a positive-displacement blower drive

The variable-frequency drive is the final controlling element, and does not measure anything.

SOLUTION 7.8

240 L/s (3840 gpm)

In International Standards units:

The span of the flow meter is (500 L/s − 100 L/s) = 400 L/s

35% of span = 0.35 × 400 L/s = 140 L/s

Therefore, when the local indicator reads 35% of span, the flow is (100 L/s + 140 L/s) = 240 L/s.

In U.S. customary units:

The span of the flow meter is (8000 gpm − 1600 gpm) = 6400 gpm

35% of span = 0.35 × 6400 gpm = 2240 gpm

Therefore, when the local indicator reads 35% of span, the flow is (1600 gpm + 2240 gpm) = 3840 gpm.

SOLUTION 7.9

e) A suspended solids sensor measuring the concentration of MLSS in the bioreactor

There is no suitable relationship between MLSS and the volume of oxygen to be supplied to a bioreactor to achieve the desired dissolved oxygen concentration.

SOLUTION 7.10

2 m upstream of the weir

A level sensor measuring flow over a weir should be positioned at a distance of at least 4 times the maximum head over the weir. In this example, the maximum head is the maximum water depth upstream of the weir minus the elevation of the weir crest, as follows:

(1.5 m − 1.0 m) = 0.5 m

Therefore, the ultrasonic level sensor should be placed at least

(4 × 0.5 m) = 2 m upstream of the weir.

SOLUTION 7.11

d) A specific chemical ion

In the case of pH, the H+ ion.

SOLUTION 7.12

c) Proportional control

Proportional control could match pump speed to maintain a certain level within the lift station or to match incoming flow if flow measurement was available.

SOLUTION 7.13

b) False

Although process instrumentation is important, some devices cannot justify the expense and upkeep of an online device. Justification of an online instrument or control system includes identifying and quantifying both tangible and intangible benefits and liabilities.

SOLUTION 7.14

b) False

Centralization and sophistication of control systems does not replace the need for periodic inspections.

SOLUTION 7.15

Instrument error is the difference between the instrument reading and the *true or actual* value.

SOLUTION 7.16

Differential head flow meters use an in-pipe constriction that produces a temporary and measurable *pressure drop* across it.

SOLUTION 7.17

d) All of the above

SOLUTION 7.18

The first meter would be purchased based on an anticipated range of 80 to 100 NTU. Once the new filtration equipment is put in place, the anticipated turbidity would be much lower. Because accuracy in this process is extremely important, a new meter should be purchased if the specifications of the original unit indicate that it is not recommended to be operated in this new lower range. In general, it is best to buy a unit to match the anticipated operating range. Selecting a device with too large a measurement span for your application may lead to inaccuracy.

SOLUTION 7.19

A reasonability check is the act of observing the indication of an instrument with process *changes* and compared with other related instrumentation.

Chapter 8
PUMPING OF WASTEWATER AND SLUDGE

Problems

PROBLEM 8.1

To control the flowrate from a positive-displacement pump, such as a progressing cavity pump or plunger pump, the operator should do which of the following?

a) Throttle the inlet (suction valve) by closing the valve as needed to control the flow.

b) Adjust the speed of rotation for a progressing cavity pump or the stroke length and/or speed for the plunger pump.

c) Install a short length of pipe smaller in diameter than the discharge pipeline.

PROBLEM 8.2

Which of the following statements is true?

a) Centrifugal pumps will pump more flow when the total head increases.

b) Positive-displacement pumps will pump approximately the same flow, regardless of the total head.

c) The flow from a positive-displacement pump will increase significantly if the total head decreases.

PROBLEM 8.3

The operators at a particular water resource recovery facility have a preference for plunger pumps over progressing cavity pumps and would like to replace the existing progressing cavity pump that conveys anaerobically digested sludge to their belt press.

a) This is a good idea because the operators prefer the plunger pump and it will simplify maintenance and training.

b) This is not recommended because plunger pumps tend to have a pulsating flow that is not desirable for a belt press, which will operate better at a more constant feed rate.

c) A vortex-type centrifugal pump would be a better choice than either a progressing cavity or plunger pump for this application.

PROBLEM 8.4

A new expansion of a water resource recovery facility has just come online and the actual flowrate is approximately 50% of the design flowrate. The facility uses an oxidation ditch-type activated sludge process with sludge wasting from the clarifier underflow. The waste activated sludge (WAS) pump has a capacity of 15.8 L/s (250 gpm). The daily WAS requirement is 190 m³/d (50 300 gpd). The WAS discharge pipeline is 150 mm (6 in.) in diameter. The design engineer recommended that the WAS pump operate intermittently; but the facility operator, from previous experience at a large facility, remembered that the WAS wasting flow should be continuous, if possible. The operator also knows that starting a motor consumes larger amounts of energy than if the motor were operating continuously. As a result, the operator would like to throttle the valve on the discharge side of the centrifugal WAS pump to reduce the flow and cause the pump to run continuously. To minimize solids accumulation in the pipeline because of settling, the velocity in the pipeline should be in the range 0.75 to 0.9 m/s (2.5 to 3 ft/sec) or more.

a) Determine the velocity in the pipeline if continuous (24-h/d) pumping occurs;

b) Determine the velocity in the pipeline with the WAS pump operating intermittently as designed; and

c) Determine the running time, assuming that the pump is to operate every 15 minutes.

PROBLEM 8.5

A progressing cavity pump is installed to convey thickened WAS from a gravity belt thickener to an anaerobic digester located approximately 150 m (500 ft) away. The operator observes that the discharge pressure is very high when the pump initially starts, but then drops off after the pump is operating for a while. The pressure, although high, is still within the design parameters of the pumping and piping system.

a) This is an abnormal operation and investigative work is needed to determine the cause of the problem.

b) This is not an unusual situation and is to be expected.

c) The pump internals need to be replaced or rebuilt.

PROBLEM 8.6

A raw wastewater wet well operates between the levels shown in the diagram. The wastewater temperature in the summer is 28 °C (82 °F) and 18 °C (65 °F) in the winter. The elevations shown are all relative to sea level. Determine the minimum net positive suction head available (NPSH available) at the pump centerline.

The following information is known:

> Hazen-Williams C_H = 140 new and C_H = 120 old
> Total length of suction pipe = 3.6 m (11.8 ft)
> Suction pipe diameter = 200 mm (8 in.)
> Pump capacity = 38.0 L/s (600 gpm)
> The total length of discharge pipe is 800 m (2625 ft)
> Discharge pipe diameter 5150 mm (6 in.)

K values for minor losses are

> Entrance loss = 0.04
> 90-deg suction elbow and 90-deg reducing elbow = 0.30, based on smaller diameter
> Gate valve fully open = 0.19
> Check valve = 0.7
> 90-deg elbow = 0.30
> 45-deg elbow = 0.20
> Exit loss = 1.00
> Saturated vapor pressure = 28.6 mm Hg (mercury) at 28 °C and 15.6 mm Hg at 18 °C
> Site barometric (atmospheric) pressure = 672 mm Hg = 89.4 kPa

The Hazen-Williams equation is as follows:

$$\text{In International Standard units: } V = 0.849 \times C_H \times R^{0.63} \times S_f^{0.54}$$

$$\text{In U.S. customary units: } V = 1.318 \times C_H \times R^{0.63} \times S_f^{0.54}$$

Where

> V = velocity, m/s (ft/sec);
> C_H = Hazen-Williams coefficient of roughness;
> R = hydraulic radius, m (ft), where R = diameter of the pipe, m (ft)/4; and
> S_f = the friction loss, m/m (ft/ft).

Head loss because of friction = $L \times S_f$, where L = pipe length, m (ft).

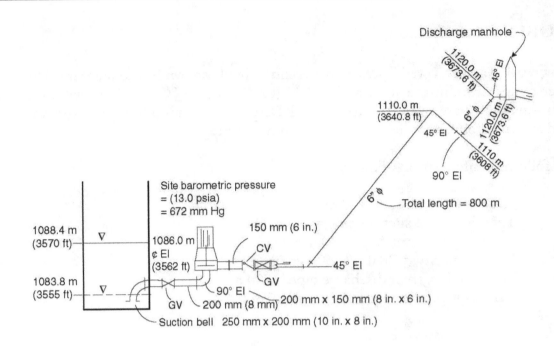

PROBLEM 8.7

The NPSH available has been determined by calculation to be 6.2 m (20.3 ft) at the design operating point on the following pump curve. The impeller diameter is 262 mm (10.3 in.).

a) Determine the NPSH required at the design operating point.

b) Is the NPSH required adequate at the design operating point?

c) Assume for this part that the pipe is new and that the operator has adjusted the wet well operating level upward to reduce the discharge head and save energy. The new operating point is at a flow of 46 L/s (729 gpm), with a total head of 21 m (68.9 ft). The NPSH available has decreased slightly to 6.1 m (20.0 ft) because of the increased flow. Does the NPSH available still exceed the NPSH required?

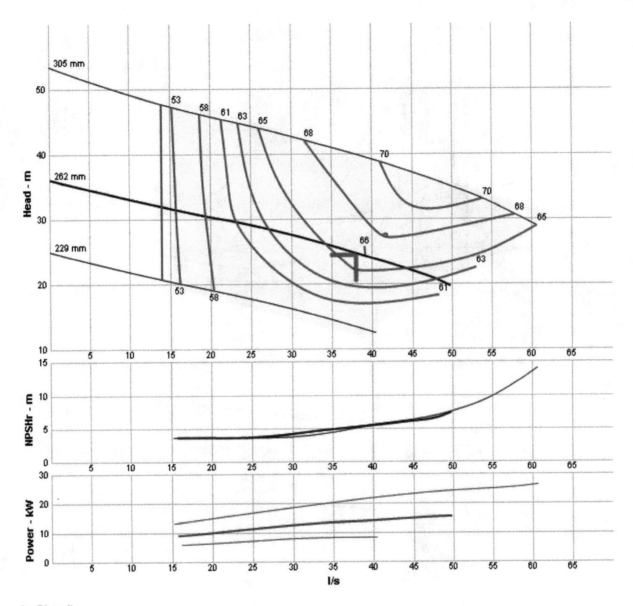

In SI units

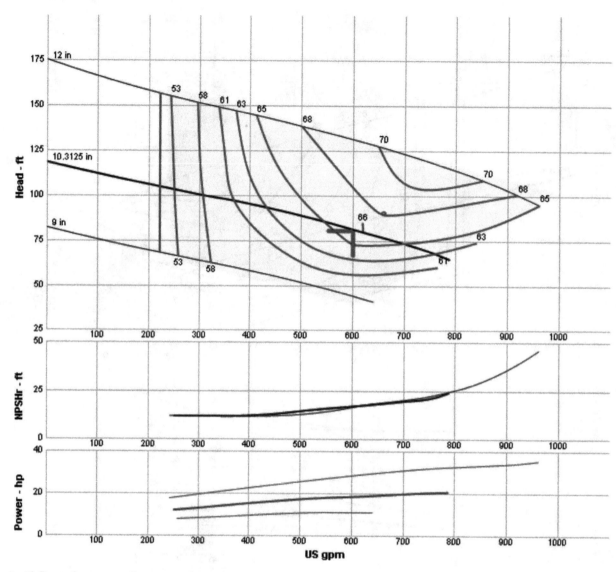

In U.S. customary units

PROBLEM 8.8

An operator wants to raise the operating levels in the wet well to reduce the pumping energy. The operator sets the maximum water level in the wet well at 3.30 m (10.8 ft) above the bottom of the wet well. The wet well is 4-m long by 2.5-m wide (13.1 ft by 8.2 ft) in plan. The pumps in the pumping station are two constant-speed pumps and have a pumping rate of 38 L/s (600 gpm). The pumps operate in a duty/automatic standby mode. They do not have an automatic alternator that would alternate the starting. The operator has set the difference between the starting and stopping water level at 0.3 m (1 ft) to minimize energy consumption. The actual minimum flow experience at the pumping station is 15 L/s (237 gpm). The centerline of the pump is 0.9 m (3 ft) above the bottom of the wet well.

   a)   Determine the minimum cycle time (i.e., the time from the start of the pump to the next time the pump starts).

b) The motor manufacturer has indicated that the pump should not start more than 5 times in one hour; otherwise, the potential exists for overheating the motor. Determine the difference in water level between stop and start required to ensure that the pump will not start more than 5 times in one hour.

(Hint: the minimum cycle time occurs when the inflow rate is 50% of the outflow rate; the derivation of this can be found in Jones et al. [2006]).

PROBLEM 8.9

For the pumping station and force main profile shown below, determine the total pumping head required for the condition of

a) The wet well at the highest level and a new pipe condition and
b) The wet well at the lowest level and an old pipe condition.

The following information is known:

> Hazen-Williams C_H = 140 new and C_H = 120 old
> Total length of suction pipe = 3.6 m (11.8 ft)
> Suction pipe diameter = 200 mm (8 in.)
> Pump capacity = 38.0 L/s (600 gpm)
> The total length of discharge pipe is 800 m (2625 ft)
> Discharge pipe diameter = 150 mm (6 in.)

K values for minor losses are

> Entrance loss = 0.04
> 90-deg suction elbow and 90-deg reducing elbow = 0.30 based on smaller diameter
> Gate valve fully open = 0.19
> Check valve = 0.7
> 90-deg elbow = 0.30
> 45-deg elbow = 0.20
> Exit loss = 1.00

Saturated vapor pressure is as follows:

> 28.6 mm Hg at 28 °C
> 15.6 mm Hg at 18 °C
> Site barometric (atmospheric) pressure 5672 mm Hg (mercury) 589.4 kPa

The Hazen-Williams equation is

$$V = 0.849 \times C_H \times R^{0.63} \times S_f^{0.54} \text{ in International Standard units}$$

$$V = 1.318 \times C_H \times R^{0.63} \times S_f^{0.54} \text{ in U.S. customary units}$$

Where

V = velocity, m/s (ft/s);

C_H = Hazen-Williams coefficient of roughness;

R = hydraulic radius, m (ft);

R = diameter of the pipe, m (ft)/4; and

S_f = the friction loss, m/m (ft/ft).

Head loss because of friction = $L \times S_f$, where L = pipe length, m (ft)

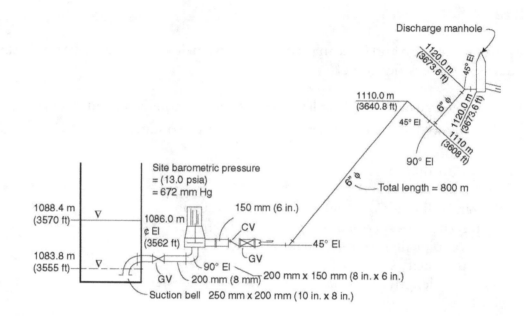

PROBLEM 8.10

An anaerobic digester currently uses a gas mixing system. The chief facility operator would like to convert the digester to a pumped mixing system using centrifugal, nonclog pumps. It is desired to "turn the digester over" every 20 minutes. The digester has a volume of 3800 m³ (1 mil. gal). Determine the capacity of the pump required to do this.

PROBLEM 8.11

The system curve for a pump is as follows:

Flow, m³/h	Total head, m	Flow, gpm	Total head, ft
0	18.3	0	60
50	19.2	220	63
100	20.7	440	68
150	23.8	660	78
200	28.0	890	92

The design pumping rate is 175 m³/h (770 gpm).

a) Plot the system curve on the attached set of performance curves;

b) Determine the head required for the pump at the design discharge;

c) Determine the appropriate impeller diameter;

d) Determine the efficiency at the operating point;

e) Calculate the motor kilowatts (horsepower) required and compare to the plot;

f) For the 200-mm (8-in.) impeller, determine the flowrate and total head at the best efficiency point; and

g) For the 200-mm (8-in.) impeller, determine the absolute maximum and minimum flowrates at which the pump should operate and provide the corresponding total heads.

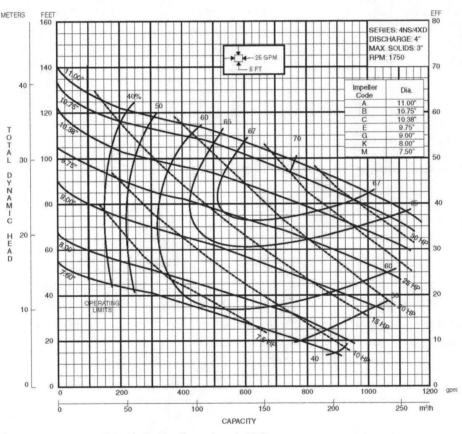

Pump curve courtesy of Goulds Pumps, Inc., a brand of ITT.

PROBLEM 8.12

The pump curves below depict the performance of a pump at 1150 rpm. For the 280-mm (11-in.) impeller shown,

a) Develop the performance curve at 875 rpm and

b) Identify the best efficiency point at 875 rpm.

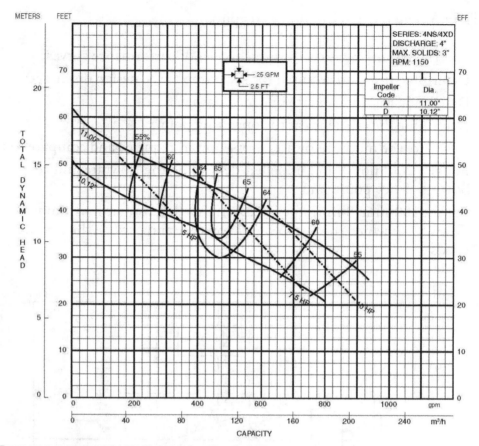

Pump curve courtesy of Goulds Pumps, Inc., a brand of ITT.

PROBLEM 8.13

Determine the head loss in 300 m (985 ft) of a 150-mm-diameter (6-in.-diameter) sludge pipe when pumping thickened primary sludge at 6% solids by weight. Ignore the effect of any fittings in the pipeline. The sludge flowrate is 16 L/s (250 gpm). Use the following figures. Determine the motor kilowatts (horsepower) assuming a wire-to-water efficiency of 0.63 and a pump efficiency of 0.70. Compare the head loss to that of water under the same conditions. Assume $C_H = 140$ for water.

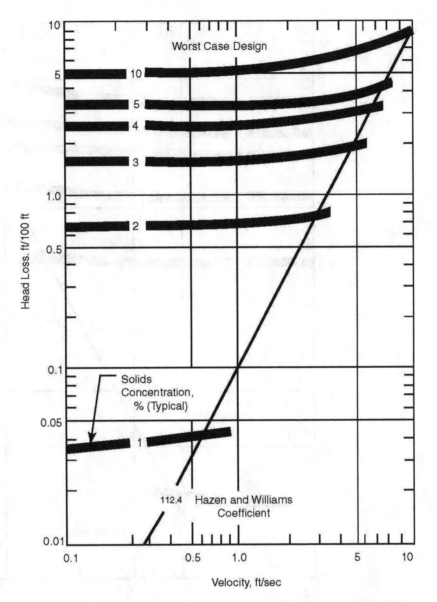

Predicted frictional head losses for worst-case design of an 6-in. diameter sludge force main (in. × 25.4 = mm; ft × 0.304 8 = m) (Mulbarger et al., 1981).

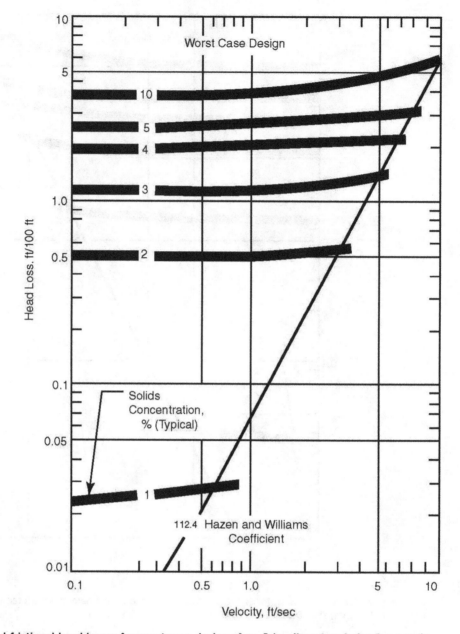

Predicted frictional head losses for worst-case design of an 8-in. diameter sludge force main (in. × 25.4 = mm; ft × 0.304 8 = m) (Mulbarger et al., 1981).

PROBLEM 8.14

An emergency has arisen that causes an operator to want to shut off a wastewater pump immediately. There are no surge-control devices on the system, only a surge anticipator valve. The surge anticipator valve has been observed to malfunction on occasion, but a service call has been placed. The wastewater pumping rate is 125 L/s (2000 gpm). The force main is 2500-m (8200-ft) long and has a diameter of 250 mm (10 in.). Determine the pressure rise that can be expected over the normal operating pressure at the pump as a result of the shutdown. Assume that ductile iron pipe is used. The pressure rise resulting from sudden instantaneous closure of a valve or emergency shutdown of a pump operating at full speed is given by the following formula:

$$\Delta h = a \times \Delta v / g$$

Where

Δh = pressure rise, or fall, m (ft);
a = elastic wave speed, m/s (ft/sec);
Δv = the change in velocity, m/s (ft/sec); and
g = gravitational constant 9.8 m/s^2 (32.2 ft/sec^2).

The wave speed varies with the type of pipe, but, for steel and ductile iron pipe, is in the range of 1100 m/s (4000 ft/sec). For polyvinyl chloride pipe, it is approximately 500 m/s (1650 ft/sec).

PROBLEM 8.15

Using the data from Problem 8.14, an operator wishes to shut off the flow in a pipeline using a mainline valve. The operator was told to close the valve slowly to avoid "water hammer". Determine the minimum length of time to close the valve to avoid excessive pressure surges. Assume that ductile iron pipe is used.

The critical, or minimum time for closure, can be determined by the following:

$$Tc = 2 \times L / a$$

Where

Tc = critical time, s;
L = length of the pipeline, m (ft); and
a = elastic wave speed, m/s (ft/sec), as defined in the previous problem.

PROBLEM 8.16

Determine the head loss in 50 m (165 ft) of 18-mm-diameter (0.75-in.-diameter) plastic pipe carrying 50% sodium hydroxide (NaOH), which has a viscosity (m) of 0.04 Pa·s (0.0008 × 5 lb-sec/sq ft). The flowrate is 0.06 L/s (1 gpm). Compare the head loss to that of pure water, which has a viscosity of approximately 0.001 Pa·s (2.08 × 10^{-5} lb-sec/sq ft). Use the Moody diagram (1944) accompanying this problem. For the solution of this problem, assume that plastic pipe has an ε/d = 0.0001. The specific gravity of 50% sodium hydroxide is 1.53.

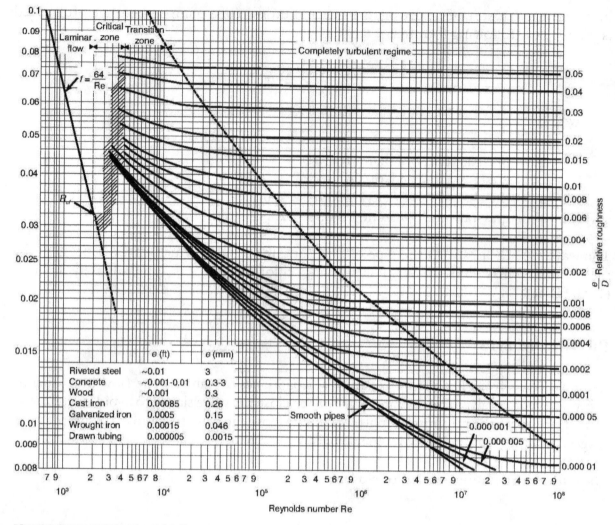

Moody diagram (Moody, 1944).

Solutions

SOLUTION 8.1

b) Adjust the speed of rotation for a progressing cavity pump or the stroke length and/or speed for the plunger pump.

To control the flow for any positive-displacement pump, either modulate the speed of rotation or the length of the stroke. For progressing cavity pumps, changing the speed of rotation is the only option; for plunger pumps or diaphragm pumps, the speed of rotation or the stroke length can be adjusted. For metering pumps, either the stroke or the frequency can be adjusted.

SOLUTION 8.2

b) Positive-displacement pumps will pump approximately the same flow, regardless of the total head.

This is why these types of pumps are the best choice when pumping sludges that vary in consistency. These pumps are also the ideal choice where constant flowrate is desired, regardless of the discharge head, for example, when pumping to a centrifuge or belt press.

SOLUTION 8.3

b) This is not recommended because plunger pumps tend to have a pulsating flow that is not desirable for a belt press, which will operate better at a more constant feed rate

Although plunger pumps are a good choice when pumping sludges, they have a pulsating discharge. When pumping to belt presses and centrifuges, it is best to have a smooth, consistent flow. Hence, a progressing cavity pump would be a better choice. A vortex centrifugal pump, although providing smooth flow, would not provide a constant discharge flowrate without a variable-speed drive and a flow control system. An air-operated diaphragm pump, although suitable for pumping sludges, has a pulsing flow similar to a plunger pump and would not be a good choice for this application.

SOLUTION 8.4

In International Standard units:

a) Determine the velocity in the pipeline if continuous (24-h/d) pumping occurs:

$$\text{Continuous flowrate} = 190 \text{ m}^3/\text{d}$$

$$V = Q/A$$

Where

V = velocity,
Q = flowrate, and
A = area of flow.

$$A = (\pi D^2)/4; A = [3.14(0.15 \text{ m})^2]/4 = 0.018 \text{ m}^2$$

$$V = (190 \text{ m}^3/\text{d})/0.018 \text{ m}^2 = 10\,560 \text{ m/d} \times 1 \text{ d}/86\,400 \text{ s} = 0.12 \text{ m/s}$$

This velocity is much lower than the recommended velocity, 0.75 to 0.9 m/s, to keep the WAS from settling out in the pipeline while pumping. Note that it will settle out when the pump is off, but will be resuspended when the pump starts again.

b) Determine the velocity in the pipeline with the WAS pump operating intermittently as designed:

$$\text{Design flowrate} = 15.8 \text{ L/s}$$

$$V = Q/A, \text{ where } V = \text{velocity}, Q = \text{flowrate}, \text{ and } A = \text{area of flow}$$

$$Q = 15.8 \text{ L/s} \times 1 \text{ m}^3/1000 \text{ L} = 0.0158 \text{ m}^3/\text{s}$$

$$A = (\pi D^2)/4; A = [3.14(0.15 \text{ m})^2]/4 = 0.018 \text{ m}^2$$

$$V = (0.0158 \text{ m}^3/\text{s})/0.018 \text{ m}^2 = 0.88 \text{ m/s}$$

This velocity is satisfactory.

c) Determine the running time assuming that the pump is to operate every 15 minutes:

Determine the operating time
Pumping rate = 0.0158 m³/s
Pumping volume = 190 m³/d
Time of operation = Volume/Pumping rate
Time of operation = (190 m³/d)/0.0158 m³/s = 12 025 s/d × 1 min/60 s = 200 min/d

If starting every 15 minutes (0.25 h), pumping duration is

Number of starts/d = 24 h/0.25 h = 96 starts/d
Duration = (200 min/d)/96 starts/d = 2.1 min/start

In U.S. customary units:

a) Determine the velocity in the pipeline if continuous (24-h/d) pumping occurs:

Continuous flowrate = 50 300 gpd
$V = Q/A$, where V = velocity, Q = flowrate, and A = area of flow
$Q = 50\ 300$ gpd × (1 cu ft/7.48 gal) × (1 d/86 400 sec) = 0.0778 cfs
$A = (\pi D^2)/4; A = [3.14(0.5 \text{ ft})^2]/4 = 0.20$ sq ft
$V = (0.0778 \text{ cfs})/0.20$ sq ft = 0.40 ft/sec

This velocity is much lower than the recommended velocity, 2.5 to 3 ft/sec, to keep the WAS from settling out in the pipeline while pumping. Note that it will settle out when the pump is off, but will be resuspended when the pump starts again.

b) Determine the velocity in the pipeline with the WAS pump operating intermittently as designed:

Design pumping rate = 250 gpm
$V = Q/A$, where V = velocity, Q = flowrate, and A = area of flow
Pump flowrate, Q = 250 gpm $\times$ 1 cu ft/7.48 gal $\times$ 1 min/60 sec = 0.56 cfs
$A = (\pi D^2)/4$; $A = [3.14(0.50 \text{ ft})^2]/4 = 0.20$ sq ft
$V = Q/A = 0.56$ cfs/0.20 sq ft = 2.8 ft/sec

This velocity is satisfactory.

c) Determine the running time assuming that the pump is to operate every 15 minutes:

Determine the operating time
Pumping rate = 250 gpm
Pumping volume = 50 300 gpd
Time of operation = Volume/Pumping rate
Time of operation = 50 300 gpd/250 gpm = 200 min/d

If starting every 15 minutes (0.25 h), pumping duration is as follows:

Number of starts/d = 24 hr/0.25 h = 96 starts/d
Duration = (200 min/d)/(96 starts/d) = 2.1 min/start

SOLUTION 8.5

b) This is not an unusual situation and is to be expected.

Thickened sludge often exhibits thixotrophic properties, which means that once it is in motion (sheared), its viscosity will decrease. This will reduce the head loss and consequently the discharge pressure at the pump. Once the flow stops, the sludge will sometimes "set up" and require significantly more pressure to make it flow again. It has been observed that some polymers, when mixed with the sludge, will cause similar problems. This is the reason that the design of pumps that handle thickened sludge should be conservatively designed.

SOLUTION 8.6

6.2 m (21.3 ft)

In International Standard units:

For this problem, it is only necessary to consider the suction side of the pump. The data for the discharge are of no concern.

NSPH available = $(P_{atm} - P_{vp})/\gamma \pm Z_w - h_L$ (Z_w is negative if pump centerline is above the suction water level.)

Minimum NPSH available occurs at minimum wet well level and maximum water temperature when the vapor pressure is maximum.

P_{atm} = 672 mm Hg (note that it is not at sea level!)

Comment: Barometric (atmospheric pressure) at sea level is 760 mm Hg. From sea level to approximately 1500 m, the pressure drops approximately 8.4 mm Hg/100 m gain in elevation. From 1500 to 3000 m, the pressure drops approximately 7.2 mm Hg/100 m gain in elevation.

Pressure in International Standard units is also expressed as Pa (Pascals) or kPa (kilopascals). Standard sea level atmospheric pressure is 101.3 kPa.

Also 1 kPa = 1 kN/m², where kN = kilonewton, the International Standard unit of force

Z_w = distance from wet well elevation to the pump volute centerline

Z_w = 1086.0 − 1083.8 = 2.4 m

g = 9.81 kN/m³ (this is the unit weight of water in the International Standard unit system)

To solve this problem, it will be necessary to determine the atmospheric pressure and the vapor pressure in terms of kPa, i.e., kN/m²

P_{atm} = 672 mm Hg × (101.3 kPa/760 mm Hg) = 89.6 kPa = 89.6 kN/m²

P_{vp} = 28.6 mm Hg × (101.3 kPa/760 mm Hg) = 3.8 kPa = 3.8 kN/m²

Calculate the head loss, h_L, in the suction pipe:

The head loss is composed of several "minor" losses and the friction loss. For minor losses, the following equation will be used: $h_L = K \times V^2/2g$

g = gravitational acceleration = 9.81 m/s²

Entrance loss into the suction bell, K = 0.04 rounded smooth:

Area = $A = (\pi D^2)/4$; $A = [3.14(0.25 \text{ m})^2]/4 = 0.049$ m²

Q = 38 L/s × (1 m³/1000 L) = 0.038 m³/s

$V = Q/A$ = (0.038 m³/s)/0.049 m² = 0.78 m/s

$V^2/2g$ = (0.78 m/s)²/(2 × 9.81 m/s²) = 0.031 m

$H_{entrance}$ = 0.04 × 0.031 m = 0.001 m

Loss in the reducing elbow 250 mm to 200 mm, K = 0.30

Use the smaller diameter for the head loss

$$\text{Area} = A = (\pi D^2)/4; \ A = [3.14(0.20 \text{ m})^2]/4 = 0.031 \text{ m}^2$$

$$V = Q/A = (0.038 \text{ m}^3/\text{s})/0.031 \text{ m}^2 = 1.22 \text{ m/s}$$

$$V^2/2 \, g = (1.22 \text{ m/s})^2/(2 \times 9.81 \text{ m/s}^2) = 0.076 \text{ m}$$

$$H_{\text{bend}} = 0.30 \times 0.076 \text{ m} = 0.023 \text{ m}$$

Gate valve loss, 200 mm diam, $K = 0.19$

$$V = 1.22 \text{ m/s}$$

$$V^2/2 \, g = (1.22 \text{ m/s})^2/(2 \times 9.81 \text{ m/s}^2) = 0.076 \text{ m}$$

$$H_{\text{gate}} = 0.19 \times 0.076 = 0.014 \text{ m}$$

Pump inlet suction elbow, 200 mm to 150 mm, $K = 0.30$

Use the smaller diameter for the head loss

$$\text{Area} = A = \pi D^2/4; \ A = [3.14(0.15 \text{ m})^2]/4 = 0.0177 \text{ m}^2$$

$$V = Q/A = (0.038 \text{ m}^3/\text{s})/0.0177 \text{ m}^2 = 2.15 \text{ m/s}$$

$$V^2/2 \, g = (2.15 \text{ m/s})^2/(2 \times 9.81 \text{ m/s}^2) = 0.236 \text{ m}$$

$$H_{\text{bend}} = 0.30 \times 0.236 \text{ m} = 0.071 \text{ m}$$

Total minor losses $= 0.001 \text{ m} + 0.023 \text{ m} + 0.014 \text{ m} + 0.071 \text{ m} = 0.109 \text{ m}$

Friction loss in the 200-mm pipe:

$$V = 1.22 \text{ m/s}$$

$$V = 0.849 \times C_{\text{H}} \times R^{0.63} \, S_f^{0.54}$$

$$R = \text{hydraulic radius} = D/4 \text{ for pipes flowing full}$$

$$R = 0.20 \text{ m}/4 = 0.05 \text{ m}$$

Substitute into the Hazen-Williams equation. Use $C_{\text{H}} = 120$ as it will generate the greatest head loss.

$$1.22 = 0.849 \times 120 \times (0.05)^{0.63} \times S_f^{0.54}$$

$$S_f^{0.54} = 1.22/[(0.849)(120)(0.151)] = 0.079$$

$$S_f = (0.079)^{1/0.54} = (0.079)^{1.85} = 0.0092 \text{ m/m}$$

$$H_{\text{friction}} = L = S_f = 3.6 \text{ m} = 0.0092 \text{ m/m} = 0.033 \text{ m}$$

Total head loss in the suction pipe = Friction loss + Total minor loss = 0.033 m + 0.109 m = 0142 m

Calculate the NPSH available:

$$\text{NPSH available} = (P_{\text{atm}} - P_{\text{vp}})/\gamma \pm Z_w - h_L$$

$$\text{NPSH available} = [(89.4 \text{ kN/m}^2 - 3.8 \text{ kN/m}^2)/9.81 \text{ kN/m}^3] - 2.4 \text{ m} - 0.142 \text{ m}$$

$$\text{NPSH available} = 6.2 \text{ m}$$

Comment: the calculation for NPSH available does not have any safety factor in it. This is discussed in a Problem 8.7. Note also that the values for minor loss K will vary from text to text. Some judgment is needed in the selection of these values. Small differences are not generally significant.

In U.S. customary units:

For this problem, it is only necessary to consider the suction side of the pump. The data for the discharge are of no concern.

NSPH available = $(P_{\text{atm}} - P_{\text{vp}})/\gamma \pm Z_w - h_L$ (Z_w is negative if pump centerline is above the suction water level.)

Minimum NPSH available occurs at minimum wet well level and maximum water temperature when the vapor pressure is maximum.

P_{atm} = 672 mm Hg (note that it is not at sea level!)

Comment: barometric (atmospheric pressure) at sea level is 760 mm Hg. From sea level to 5000 ft, the pressure drops approximately 25.6 mm Hg/1000 ft gain in elevation. From 5000 to 10 000 ft, the pressure drops approximately 21.9 mm Hg/1000 ft gain in elevation. The pressure at sea level is 14.7 psi.

Pressure in U.S. customary units is typically expressed as psi (lb/sq in.) or lb/sq ft.

Z_w = distance from wet well elevation to the pump volute centerline

Z_w = 3562 − 3555 = 7.0 ft

γ = 62.4 lb/cu ft (the unit weight of water in U.S. customary units)

To solve this problem, it will be necessary to determine the atmospheric pressure and the vapor pressure in terms of lb/sq ft.

P_{atm} = 672 mm Hg × (14.7 lb/sq in.)/760 mm Hg = 13.0 lb/sq in. × 144 sq in./sq ft = 1872 lb/sq ft

P_{vp} = 28.6 mm Hg × (14.7 lb/sq in.)/760 mm Hg = 0.55 lb/sq in. × 144 sq in./sq ft = 79.2 lb/sq ft

The head loss is composed of several "minor" losses and the friction loss. For minor losses the following equation will be used:

$$h_L = K \times V^2/2\,g$$

g = gravitational acceleration = 32.2 ft/sec²

Entrance loss into the suction bell, K = 0.04 rounded smooth:

Area = $A = (\pi D^2)/4$; $A = [3.14(0.83 \text{ ft})^2]/4$ = 0.54 sq ft

Q = 600 gpm × (1 cu ft/7.48 gal) × (1 min/60 sec) = 1.34 cfs

$V = Q/A$ = 1.34 cfs/0.54 sq ft = 2.48 ft/sec

$V^2/2\,g$ = (2.48 ft/sec)²/(2 × 32.2 ft/sec²) = 0.095 ft

$H_{entrance}$ = 0.04 × 0.095 ft = 0.004 ft

Loss in the reducing elbow 10 in. to 8 in., K = 0.30:

Use the smaller diameter for the head loss

Area = $A = (\pi D^2)/4$; $A = [3.14(0.67 \text{ ft})^2]/4$ = 0.35 sq ft

$V = Q/A$ = 1.34 cfs/0.35 sq ft = 3.8 ft/sec

$V^2/2\,g$ = (3.8 ft/sec)²/(2 × 32.2 ft/sec²) = 0.22 ft

H_{bend} = 0.30 × 0.22 ft = 0.07 ft

Gate valve loss, 8 in. diam, K = 0.19

V = 3.8 ft/sec

$V^2/2\,g$ = 0.22 ft

H_{gate} = 0.19 × 0.22 = 0.04 ft

Pump inlet suction elbow 8 in. to 6 in., K = 0.30

Use the smaller diameter for the head loss

$$\text{Area} = A = (\pi D^2)/4; A = [3.14(0.5 \text{ ft})^2]/4 = 0.20 \text{ sq ft}$$

$$V = Q/A = 1.34 \text{ cfs}/0.20 \text{ sq ft} = 6.7 \text{ ft/sec}$$

$$V^2/2g = (6.7 \text{ ft/sec})^2/(2 \times 32.2 \text{ ft/sec}^2) = 0.70 \text{ ft}$$

$$H_{\text{bend}} = 0.30 \text{ ft} \times 0.70 \text{ ft} = 0.21 \text{ ft}$$

$$\text{Total minor losses} = 0.004 \text{ ft} + 0.07 \text{ ft} + 0.04 \text{ ft} + 0.21 \text{ ft} = 0.32 \text{ ft}$$

Friction loss in the 8-in. pipe:

$$V = 3.8 \text{ ft/sec}$$

$$V = 1.318 \times C_H \times R^{0.63} S_f^{0.54}$$

$$R = \text{hydraulic radius} = D/4 \text{ for pipes flowing full}$$

$$R = 0.67 \text{ ft}/4 = 0.17 \text{ ft}$$

Substitute into the Hazen-Williams equation. Use $C_H = 120$ as it will generate the greatest head loss.

$$3.8 = 1.318 \times 120 \times (0.17)^{0.63} \times S_f^{0.54}$$

$$S_f^{0.54} = 3.8/[(1.318)(120)(0.33)] = 0.073$$

$$S_f = (0.073)^{1/0.54} = (0.073)^{1.85} = 0.008 \text{ ft/ft}$$

Comment: This should be identical to that of the International Standard units, but rounding errors preclude this.

$$H_{\text{friction}} = L \times S_f = 11.8 \text{ ft} \times 0.008 \text{ ft/ft} = 0.09 \text{ ft}$$

Total head loss in the suction pipe = Friction loss + Total minor loss = 0.32 ft + 0.09 ft = 0.41 ft

Calculate the NPSH available:

$$\text{NPSH available} = (P_{\text{atm}} - P_{\text{vp}})/\gamma \pm Z_w - h_L$$

$$\text{NPSH available} = [(1872 \text{ lb/sq ft} - 79.2 \text{ lb/sq ft})/62.4 \text{ lb/cu ft}] - 7.0 \text{ ft} - 0.41 \text{ ft}$$

$$\text{NPSH available} = 21.3 \text{ ft}$$

Comment: The calculation for NPSH available does not have any safety factor in it. This is discussed in Problem 8.7. Note also the values for minor loss K will vary from text to text.

Some judgment is needed in the selection of these values. Small differences are not generally significant.

SOLUTION 8.7

In International Standard units:

a) The operating point is 38 L/s at 24.5-m head. The impeller diameter is shown to be 262 mm in diameter. From the NPSH required curve, the NPSH required is 5 m.

b) Because 6.2 m available is greater than the 5 m required, the pump operation should be satisfactory. However, there is no safety factor or safety margin for NPSH. It is recommended that the NPSH available be decreased by 1.5 m (5 ft) or that NPSH required by the manufacturer be increased by 35% (Jones et al., 2006). Using this criterion, the NPSH available with the safety factor should be 6.2 m − 1.5 m = 4.7 m. This is less than the 5 m required by the manufacturer. Alternatively, the NPSH required should be 5 m × 1.35 = 6.75 m. Because only 6.2 m is available, the NPSH available is inadequate by this criterion also. In summary, the NPSH available is inadequate for satisfactory pump operation.

c) The new operating point is 46.5 L/s and 21.5 m head. The NPSH required as determined from the curve is 6 m. The NPSH available with the safety factor is 6.1 m + 1.5 m = 4.6 m. This is less than the 6 m required by the manufacturer. Alternatively, the NPSH required should be 6 m × 1.35 = 8.1 m. Because only 6.1 m is available, the NPSH available is inadequate by this criterion also. In summary, the NPSH available is inadequate for satisfactory pump operation.

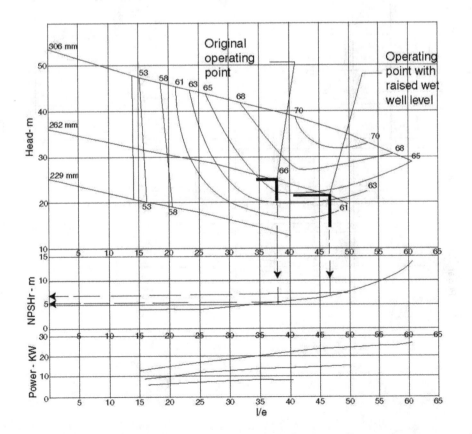

In U.S. customary units:

a) The operating point is 600 gpm at 80-ft head. The impeller diameter is shown to be 10.3 in. From the NPSH required curve, the NPSH required is 16.4 ft.

b) Because the 20.3 ft available is greater than 16.4 ft required, the pump operation should be satisfactory. However, there is no safety factor or safety margin for NPSH. It is recommended that the NPSH available be decreased by 1.5 m (5 ft) or the NPSH required by the manufacturer be increased by 35% (Jones et al., 2006). Using this criterion, the NPSH available with the safety factor should be 20.3 ft + 5 ft = 15.3 ft. This is less than the 16.4 ft required by the manufacturer. Alternatively, the NPSH required should be 16.4 ft × 1.35 = 22.1 ft. Because only 20.3 ft is available, the NPSH available is inadequate by these criteria also. In summary, the NPSH available is inadequate for satisfactory pump operation.

c) The new operating point is 729 gpm and 68.9-ft head. The NPSH required as determined from the curve is 19.7 ft. The NPSH available with the safety factor is 20 ft + 5 ft = 15 ft. This is less than the 19.7 ft required by the manufacturer. Alternatively, the NPSH required should be 19.7 ft × 1.35 = 26.6 ft. Because only 20 ft is available, the NPSH available is inadequate by these criteria also. In summary, the NPSH available is inadequate for satisfactory pump operation.

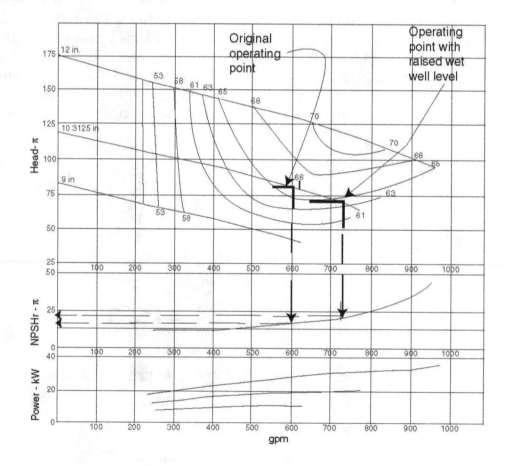

SOLUTION 8.8

In International Standard units:

a) The cycle time is time from the start of operation until that same pump restarts. It comprises the time to fill the wet well from pump stop to pump start plus the time to empty the wet well to the point the pump stops again.

Time to fill wet well = T_f

Time to empty wet well = T_e

Cycle time = $T_f + T_e$, where T_f = volume of wet well/Q_{in} (Q_{in} = inflow rate) and T_e = volume of the wet well/($Q_{out} - Q_{in}$) where Q_{out} = outflow rate

Volume of the wet well between start and stop levels = 4 m × 2.5 m × 0.3 m = 3 m³

$Q_{pump} = Q_{out} = 38$ L/s × (1 m³/1000 L) × 60 s/min = 2.28 m³/min

Minimum cycle time occurs when the inflow rate equals 50% of the outflow rate.

50% of the inflow rate = 1.14 m³/min

T_f = 3 m³/1.14 m³/min = 2.63 min

T_e = 3 m³/(2.28 − 1.14) m³/min = 2.63 min

Cycle time = 2.63 min + 2.63 min = 5.3 min

Check the cycle time at the actual minimum flow = 15 L/s × 60 s/min × 1 m³/1000 L = 0.9 m³/min

T_f = 3 m³/0.9 m³/min = 3.33 min

T_e = 3 m³/(2.28 − 0.9) m³/min = 2.17 min

Cycle time at actual minimum flow = 3.33 min + 2.17 min = 5.5 min

Comment: this demonstrates that the minimum cycle time does not always occur at the minimum flowrate!

b) Determine the cycle time for no more than five starts/hour

5 starts/h × 1 h/60 min = 1 start/12 min; cycle time = 12 min

T_f = volume of the wet well/(1.14 m³/min)

T_e = volume of the wet well/(2.28 − 1.14) m³/min

T_e = volume of the wet well/(1.14 m³/min)

Cycle time = $T_f + T_e$

Let V_{ww} = volume of the wet well from stop to start of the pump

Cycle time = V_{ww}/(1.14 m³/min) + V_{ww}/(1.14 m³/min)

Cycle time = 2 × [V_{ww}/(1.14 m³/min)]

V_{ww} = (12 min × 1.14 m³/min)/2

V_{ww} = 6.84 m³

Difference in water levels from stop to start = 6.84 m³/(4 m × 2.5 m) = 0.68 m

The stop water level would be 3.3 m − 0.68 m = 2.62 m above the bottom of the wet well. This is above the pump centerline, so is probably satisfactory from an NPSH standpoint, but this should be checked. See problem 8.6 above.

Comment: minimum cycle time = 2 × [V_{ww}/(1.14 m³/min)], but 1.14 m³/min = Q_{pump}/2. So, we can rewrite this as

$$\text{Minimum cycle time} = 4 \times V_{ww}/Q_{pump}$$

or

$$V_{ww} = Q_{pump} \times \text{Cycle time}/4$$

Note that it is possible to double the calculated cycle time by providing automatic alternation of the lead and lag pumps.

In U.S. customary units:

a)   The cycle time is the time from the start of operation until that same pump restarts. It comprises the time to fill the wet well from pump stop to pump start plus the time to empty the wet well to the point the pump stops again.

T_f = Volume of wet well/Q_{in}, where Q_{in} = inflow rate

T_e = Volume of the wet well/($Q_{out} − Q_{in}$), where Q_{out} = outflow rate

Volume of the wet well between start and stop levels = 13.1 ft × 8.2 ft × 1 ft = 107.42 cu ft × 7.48 gal/cu ft = 803 gal

$Q_{pump} = Q_{out} = 600$ gpm

Minimum cycle time occurs when the inflow rate is 50% of the outflow rate.

50% of the inflow rate = 300 gpm

$T_f = 803$ gal/300 gpm = 2.68 min

$T_e = 803$ gal/(600 − 300) gpm = 2.68 min

Cycle time = 2.68 min + 2.68 min = 5.4 min

Check the cycle time at the actual minimum flow = 237 gpm

$T_f = 803$ gal/237 gpm = 3.3 min

$T_e = 803$ gal/(600 − 237) gpm = 2.21 min

Cycle time at actual minimum flow = 3.3 min + 2.2 min = 5.5 min

Comment: This demonstrates that the minimum cycle time does not always occur at the minimum flowrate!

b)   Determine the cycle time for no more than five starts per hour

5 start/hr × 1 hr/60 min = 1 start/12 min; cycle time = 12 min

T_f = volume of the wet well/(300 gpm)

T_e = volume of the wet well/(600 − 300 gpm)

T_e = volume of the wet well/(300 gpm)

Let V_{ww} = volume of the wet well from stop to start of the pump

Cycle time = $T_f + T_e$

Cycle time = V_{ww}/(300 gpm) + V_{ww}/(300 gpm)

Cycle time = 2 × [V_{ww}/(300 gpm)]

V_{ww} = 12 min × 300 gpm/2

V_{ww} = 1800 gal × 1 cu ft/7.48 gal = 241 cu ft

Difference in water levels from stop to start = 241 cu ft/(13.1 ft × 8.2 ft) = 2.24 ft

Comment: minimum cycle time = $2 \times [V_{ww}/(300 \text{ gpm})]$, but 300 gpm = $Q_{pump}/2$

So, we can rewrite this as

$$\text{Minimum cycle time} = 4 \times V_{ww}/Q_{pump}$$

or

$$V_{ww} = Q_{pump} \times \text{Cycle time}/4$$

Note that it is possible to double the calculated cycle time by providing automatic alternation of the lead and lag pumps.

SOLUTION 8.9

In International Standard units:

a) Wet well level at its highest elevation, new pipe $C_H = 140$

Calculate head loss in the suction pipe.

Entrance loss into the suction bell, $K = 0.04$ rounded smooth:

Area = $A = (\pi D^2)/4$; $A = [3.14(0.25 \text{ m})^2]/4 = 0.049 \text{ m}^2$

$Q = 38 \text{ L/s} \times (1 \text{ m}^3/1000 \text{ L}) = 0.038 \text{ m}^3/\text{s}$

$V = Q/A = 0.038 \text{ m}^3/\text{s} \div 0.049 \text{ m}^2 = 0.78 \text{ m/s}$

$V^2/2g = (0.78 \text{ m/s})^2/(2 \times 9.81 \text{ m/s}^2) = 0.031 \text{ m}$

$H_{entrance} = 0.04 \times 0.031 \text{ m} = 0.001 \text{ m}$

Loss in the reducing elbow 250 mm to 200 mm, $K = 0.30$:

Use the smaller diameter for the head loss

Area = $A = (\pi D^2)/4$; $A = [3.14(0.20 \text{ m})^2]/4 = 0.031 \text{ m}^2$

$V = Q/A = 0.038 \text{ m}^3/\text{s} \div 0.031 \text{ m}^2 = 1.22 \text{ m/s}$

$V^2/2g = (1.22 \text{ m/s})^2/(2 \times 9.81 \text{ m/s}^2) = 0.076 \text{ m}$

$H_{bend} = 0.30 \times 0.076 \text{ m} = 0.023 \text{ m}$

Gate valve loss, 200 mm diam, $K = 0.19$

$V = 1.22$ m/s

$V^2/2g = (1.22 \text{ m/s})^2/(2 \times 9.81 \text{ m/s}^2) = 0.076$ m

$H_{\text{gate}} = 0.19 \times 0.076 \text{ m} = 0.014$ m

Pump inlet suction elbow, 200 mm to 150 mm, $K = 0.30$

Use the smaller diameter for the head loss

Area $= A = (\pi D^2)/4$; $A = [3.14(0.15 \text{ m})^2]/4 = 0.0177 \text{ m}^2$

$V = Q/A = 0.038 \text{ m}^3/\text{s} \div 0.0177 \text{ m}^2 = 2.15$ m/s

$V^2/2g = (2.15 \text{ m/s})^2/(2 \times 9.81 \text{ m/s}^2) = 0.236$ m

$H_{\text{bend}} = 0.30 \times 0.236 \text{ m} = 0.071$ m

Total minor losses $= 0.001 \text{ m} + 0.023 \text{ m} + 0.014 \text{ m} + 0.071 \text{ m} = 0.109$ m

Friction loss in the 200-mm pipe:

$V = 1.22$ m/s

$V = 0.849 \times C_H \times R^{0.63} S_f^{0.54}$

$R =$ hydraulic radius $= D/4$ for pipes flowing full

$R = 0.20 \text{ m}/4 = 0.05$ m

Substituting into the Hazen-Williams equation:

$1.22 = 0.849 \times 140 \times (0.05)^{0.63} \times S_f^{0.54}$

$S_f^{0.54} = 1.22/[(0.849)(140)(0.151)] = 0.068$

$S_f = (0.068)^{1/0.54} = (0.068)^{1.85} = 0.0069$ m/m

$H_{\text{friction}} = L \times S_f = 3.6 \text{ m} \times 0.0069 \text{ m/m} = 0.025$ m

Elevation of the energy grade line at the pump suction $=$ Wet well elevation $- H_{\text{minor}} = H_{\text{friction}} = 1088.4 \text{ m} - 0.109 \text{ m} - 0.025 \text{ m} = 1088.27$ m at the pump suction

Head loss in the 150-mm discharge pipe:

Area $= A = (\pi D^2)/4$; $A = [3.14(0.15 \text{ m})^2]/4 = 0.018 \text{ m}^2$

$Q = 38$ L/s $\times$ (1 m³/1000 L) $= 0.038$ m³/s

$V = Q/A = 0.038$ m³/s $\div 0.018$ m² $= 2.1$ m/s

$V^2/2g = (2.1$ m/s$)^2/(2 \times 9.81$ m/s²$) = 0.22$ m

Because the discharge diameter and fittings are the same diameter, we can sum the minor loss K values to save time:

Check valve $K = 0.7$

Gate valve $K = 0.19$

Three at 45-deg elbows $K = 0.20$ each, total $= 3 \times 0.20 = 0.60$

90-deg elbow, $K = 0.30$

Exit loss, $K = 1.0$

Total $K = 0.70 + 0.19 + 0.60 + 0.30 + 1.00 = 2.79$

$H_{\text{minor}} = KV^2/2\,g = 2.79 \times (0.22$ m$) = 0.61$ m

Friction loss in 150-mm discharge pipe:

$V = 0.849 \times C_H \times R^{0.63} S_f^{0.54}$

$R =$ hydraulic radius $= D/4$ for pipes flowing full

$R = 0.15$ m/4 $= 0.0375$ m

Substituting into the Hazen-Williams equation:

$2.1 = 0.849 \times 140 \times (0.0375)^{0.63} \times S_f^{0.54}$

$S_f^{0.54} = 2.1/[(0.849)(140)(0.126)] = 0.140$

$S_f = (0.140)^{1/0.54} = (0.140)^{1.85} = 0.026$ m/m

$H_{\text{friction}} = L \times S_f = 800$ m $\times 0.026$ m/m $= 20.8$ m

Elevation of the energy grade line at the pump discharge = Pump discharge elevation + $H_{\text{minor}} + H_{\text{friction}} = 1120.0$ m $+ 0.61$ m $+ 20.8$ m $= 1141.4$ m at the pump discharge

Total pumping head = Elevation of energy grade line at pump discharge − Elevation of energy grade line at the pump suction $= 1141.4$ m $- 1088.27$ m $= 53.13$ m

Note the pump head can also be determined at the static elevation difference plus the friction and minor losses in both the suction and the discharge piping as follows:

Static head $= 1120.0$ m $- 1088.4$ m $= 31.6$ m

Friction and minor losses $= 0.109$ m $+ 0.025$ m $+ 0.61$ m $+ 20.8$ m $= 21.54$ m

Total pumping head $= 21.54$ m $+ 31.6$ m $= 53.14$ m

b)   Wet well level at its lowest position and old pipe $C_H = 120$

Because the flowrate has not changed, the minor losses in the suction pipe and discharge pipe will not change. They are determined by $K \times V^2/2\, g$; as long as the velocity does not change, the minor losses will not change. So, there is no need to recalculate them.

Minor losses in the suction pipe $= 0.109$ m

Friction loss in the 200-mm pipe:

$V = 1.22$ m/s

$V = 0.849 \times C_H \times R^{0.63}\, S_f^{0.54}$

$R =$ hydraulic radius $= D/4$ for pipes flowing full

$R = 0.20$ m$/4 = 0.05$ m

Substituting into the Hazen-Williams equation:

$1.22 = 0.849 \times 120 \times (0.05)^{0.63} \times S_f^{0.54}$

$S_f^{0.54} = 1.22/[(0.849)(120)(0.151)] = 0.079$

$S_f = (0.079)^{1/0.54} = (0.079)^{1.85} = 0.0092$ m/m

$H_{friction} = L \times S_f = 3.6$ m $\times 0.0092$ m/m $= 0.033$ m

Minor losses in the discharge pipe $= 0.61$ m

Friction loss in 150-mm discharge pipe:

$V = 0.849 \times C_H \times R^{0.63}\, S_f^{0.54}$

$R =$ hydraulic radius $= D/4$ for pipes flowing full

$R = 0.15$ m$/4 = 0.0375$ m

Substituting into the Hazen-Williams equation:

$$2.1 = 0.849 \times 120 \times (0.0375)^{0.63} \times S_f^{0.54}$$

$$S_f^{0.54} = 2.1/[(0.849)(120)(0.126)] = 0.164$$

$$S_f = (0.164)^{1/0.54} = (0.164)^{1.85} = 0.035 \text{ m/m}$$

$$H_{\text{friction}} = L \times S_f = 800 \text{ m} \times 0.035 \text{ m/m} = 28.0 \text{ m}$$

Elevation of the energy grade line at the pump suction = Wet well elevation − H_{minor} = H_{friction} = 1083.8 m − 0.109 m − 0.033 m = 1083.66 m at the pump suction

Elevation of the energy grade line at the pump discharge = Pump discharge elevation + H_{minor} + H_{friction} = 1120.0 m + 0.61 m + 28.0 m = 1148.6 m at the pump discharge

Total pumping head = 1148.6 m = 1083.66 m = 64.94 m

Or, alternatively,

Static head = 1120.0 m − 1083.8 m = 36.2 m

Friction and minor losses = 0.109 m + 0.033 m + 0.61 m + 28.0 m = 28.75 m

Total pumping head = 36.2 m + 28.75 m = 64.95 m

Comment: These two conditions, "a" and "b", typically represent the extremes in pumping head. Condition a is the minimum head; condition b the maximum head.

In U.S. customary units:

The solution in U.S. customary units is quite similar. The reader is encouraged to review Solution 8.6 for the methodology of determining the pipe losses using U.S. customary units.

SOLUTION 8.10

11 400 m³/h (50 000 gpm)

In metric units:

Pumping rate = Volume/Turnover time

Pumping rate = 3800 m³/20 min = 190 m³/min × 60 min/h = 11 400 m³/h

In U.S. customary units:

Pumping rate = Volume/Turnover time

Pumping rate = 1 mil. gal/20 min = 50 000 gpm

SOLUTION 8.11

In International Standard units:

a)   See the figure below for a plot of the system curve.

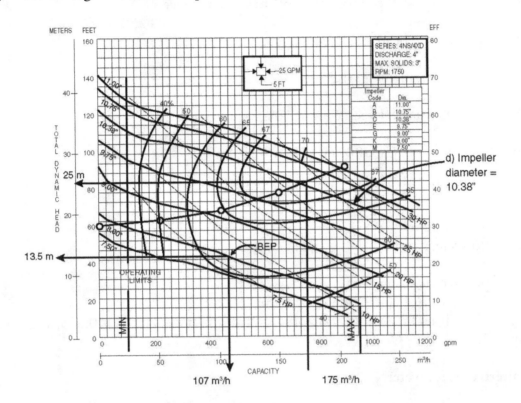

b)   At 175 m³/h; total head = 25 m

c)   Impeller diameter = 260 mm (10.38 in.)

d)   Efficiency lies between 67 and 70%, say 68.5%

e)   kW = (9.81 kN/m³) × (Q, m³/s) × (H, m)/ep kW = [(9.81 kN/m³) × (175 m³/h × 1 h/3600 s) × 25 m]/0.685 = 17.4

f)   From plot, best efficiency point (BEP) = 107 m³/h at 13.5 m head

g)   Minimum flowrate is approximately 25 m³/h; maximum is approximately 215 m³/h

General comment: A pump should never be operated at a flowrate greater than the performance curve provided by the pump manufacturer for the given impeller. This condition is called "runout". Also, a pump should never be operated at less flow than that for which the manufacturer has provided efficiency data. In the figure below, the pump should not be operated at a flowrate less than approximately 25 L/s. For pumps with specific speeds less than 5200 (International Standard units) or 4500 (U.S. customary units), the preferred operating range is 70 to 120% of the flow at the BEP. For higher specific speeds, the preferred operating range is 75 to 115% of the flow at the BEP (Jones et al., 2006).

In U.S. customary units:

a) See the figure above for a plot of the system curve

b) At 770 gpm, head = 84 ft

c) Impeller diameter = 10.38 in.

d) Efficiency lies between 67 and 70%, say 68.5%

e) hp = $\gamma \times Q \times H/550 \times$ ep

Where

γ = 62.4 lb/cu ft;
Q = flowrate, cfs;
H = head, ft;
ep = pump efficiency;
Q = 770 gpm $\times$ 1 cu ft/7.48 gal $\times$ 1 min/60 s = 1.72 cfs;
hp = (62.4 lb/cu ft) $\times$ (1.72 cfs) $\times$ (84 ft)/(550 ft-lb/sec $\times$ hp) $\times$ 0.685; and
hp = 23.8.

As an alternative for water:

$$hp = Q \times H/3960 \times ep$$

Where

Q = flowrate, gpm;
H = head, ft; and
ep = pump efficiency.

Comment: This form is often more convenient to use. The denominator is actually 3955.8; frequently, 3960 is used for expediency. Little error is introduced.

$$hp = \frac{(770 \text{ gpm}) \times (84 \text{ ft})}{(3960 \times 0.685)} = 23.8$$

f) From the plot, BEP = 470 gpm at 4 ft head

g) Minimum flowrate is approximately 110 gpm; maximum is approximately 945 gpm

SOLUTION 8.12

Pick the $Q = 0$ condition and then arbitrarily select points on the performance curve to give a good distribution of points to allow drawing a smooth curve. It is advantageous to pick the points where the efficiency is known because the efficiency "follows" the head and discharge points. The affinity laws will be used in the solution.

In International Standard units:

a) $\varpi_2/\varpi_1 = Q_2/Q_1$ and $(\varpi_2/\varpi_1)^2 = H_2/H_1$

At $Q = 0$, head = 19.5 m

$\varpi_1 = 1150$ rpm; $\varpi_2 = 875$ rpm; $Q_1 = 0$

$875/1150 = Q_1/0; Q_1 = 0$

$(875/1150)^2 = H_2/19.5$ m; $H_2 = 19.5 \times 0.58 = 11.3$ m

Therefore, $Q = 0$, $H = 11.3$ m is a point on the new performance curve

At $Q = 45$ m³/h, head = 15.9 m

$875/1150 = Q_1/45$ m³/h; $Q_1 = 34.2$ m³/h

$(875/1150)^2 = H_2/15.9$ m; $H_2 = 15.9 \times 0.58 = 9.2$ m

Therefore, $Q = 34.2$ m³/h, $H = 9.2$ m is a point on the new performance curve. The efficiency for this new point remains the same at 55%.

In similar fashion, the other points on the performance curve are determined.

n_2	875 rpm							
n_1	1150 rpm							
Q_1	0	45	90	113	134	170	200	m³/h
H_1	19	15.9	14	13	12	10.4	8.5	m
ep		0.55	0.64	0.655	0.64	0.6	0.55	
Q_2	0	34.2	68.5	86.0	102.0	129.3	152.2	m³/h
H_2	11.0	9.2	8.1	7.5	6.9	6.0	4.9	m
ep		0.55	0.64	0.655	0.64	0.6	0.55	

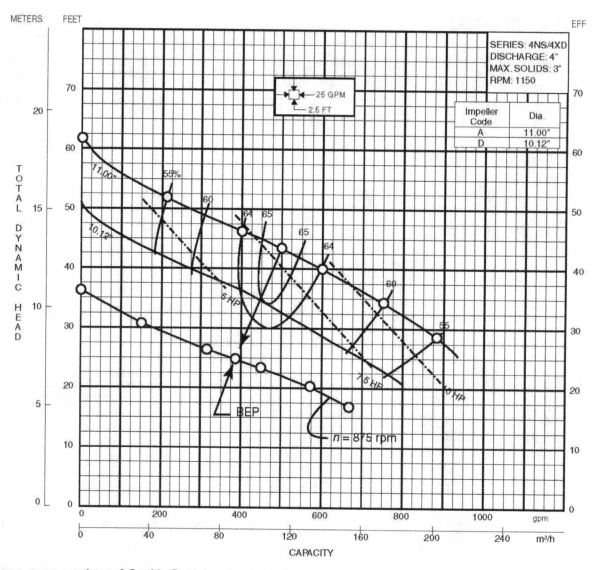

Pump curve courtesy of Goulds Pumps, a brand of ITT.

b) At the BEP, ep = 0.655, Q = 86.0 m³/h, and H = 7.5 m

In U.S. customary units:

MO(a) $\varpi_2/\varpi_1 = Q_2/Q_1$ and $(\varpi_2/\varpi_1)^2 = H_2/H_1$

At Q = 0, head = 62 ft

ϖ_1 = 1150 rpm; ϖ_2 = 875 rpm; Q_1 = 0

875/1150 = Q1/0; Q1 = 0

$(875/1150)^2 = H_2/62$ ft; H_2 = 62 ft × 0.58 = 35.9 ft

Therefore, $Q = 0$, $H = 35.9$ ft is a point on the new performance curve

At $Q = 200$ gpm, head $= 52$ ft

$875/1150 = Q_1/200$ gpm; $Q_1 = 152$ gpm

$(875/1150)^2 = H_2/52$ ft; $H_2 = 52$ ft $\times 0.58 \times 30.1$ ft

Therefore, $Q = 152$ gpm, $H = 30.1$ ft is a point on the new performance curve. The efficiency for this new point remains the same at 55%.

In similar fashion, the other points on the performance curve are determined.

n_2	875 rpm							
n_1	1150 rpm							
Q_1	0	200	400	500	590	750	880	gpm
H_1	62	52	46	43	40	34	28	ft
ep		0.55	0.64	0.655	0.64	0.6	0.55	
Q_2	0	152	304	380	449	571	697	gpm
H_2	35.9	30.1	26.6	24.9	23.2	19.7	16.2	ft
ep		0.55	0.64	0.655	0.64	0.6	0.55	

b) At the BEP, ep $= 0.655$, $Q = 379$ gpm, and $H = 2.3$ ft.

SOLUTION 8.13

The plot was obtained from Mulbarger et al. (1981) and is only available in U.S. customary units.

In International Standard units:

$Q = 16$ L/s $\times$ (1 m^3/1000 L) $= 0.016$ m^3/s

Diameter $= 150$ mm

Area $= A = (\pi D^2)/4 = [3.14(0.15 \text{ m})^2]/4 = 0.018$ m^2

$V = Q/A = 0.016$ m^3/s $\div 0.018$ m$^2 = 0.89$ m/s

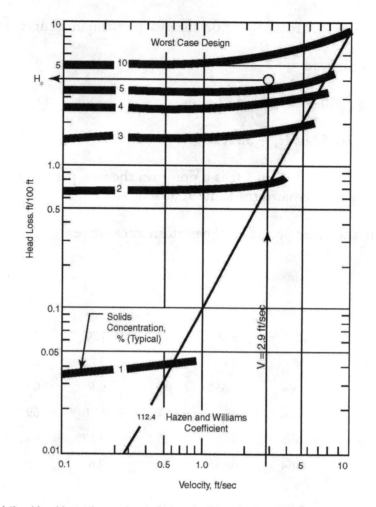

Predicted frictional head losses for worst-case design of a 6-in. diameter sludge force main (in. × 25.4 = mm; ft × 0.304 8 = m) (Mulbarger et al., 1981).

From the figure shown above,

$$S_f = 4 \text{ ft}/100 \text{ ft} = 4 \text{ m}/100 \text{ m}$$

$$H_L = S_f \times L = (4 \text{ m}/100 \text{ m}) \times 300 \text{ m} = 12 \text{ m}$$

$$\text{Motor kW} = (9.81 \text{ kN/m}^3) \times (Q, \text{m}^3/\text{s}) \times (H, \text{m})/ep$$

$$\text{kW} = (9.81 \text{ kN/m}^3) \times (0.016 \text{ m}^3/\text{s}) \times (12 \text{ m})/0.70 = 2.7 \text{ kW}$$

Note that the wire-to-water efficiency is not used here because we are only trying to find the motor output kilowatts not the input electrical energy. Also we will assume that the sludge has the same unit weight as water. The unit weight could easily be adjusted if necessary, if it were known. Just multiply the 9.81 kN/m³ by the specific gravity of the sludge.

Determine the H_L for water under the same conditions.

$$V = 0.849 \times C_H R^{0.63} \times S_f^{0.54}$$

$$R = \text{hydraulic radius} = D/4 \text{ for pipes flowing full} = 015 \text{ m}/4 = 0.0375 \text{ m}$$

$$R^{0.63} = (0.0375)^{0.63} = 0.126 \text{ m}$$

Substituting into the Hazen-Williams equation and using $C_H = 140$:

$$0.89 = 0.849 \times (140) \times (0.126) \times S_f^{0.54}$$

$$S_f^{0.54} = 0.89/(0.849 \times 140 \times 0.126) = 0.059$$

$$S_f = (0.059)^{1/0.54} = (0.059)^{1.85}$$

$$S_f = 0.0054 \text{ m/m}$$

$$H_L = S_f \times L = 0.0054 \text{ m/m} \times 300 \text{ m} = 1.6 \text{ m}$$

In U.S. customary units:

$$Q = 250 \text{ gpm} \times 1 \text{ cu ft}/7.48 \text{ gal} \times 1 \text{ min}/60 \text{ sec} = 0.56 \text{ cfs}$$

$$\text{Diameter} = 6 \text{ in.} = 0.5 \text{ ft}$$

$$\text{Area} = A = (\pi D^2)/4 = [3.14(0.5 \text{ ft})^2]/4 = 0.20 \text{ sq ft}$$

$$V = Q/A = 0.56 \text{ cfs}/0.20 \text{ sq ft} = 2.8 \text{ ft/sec}$$

From the figure shown above:

$$S_f = 4 \text{ ft}/100 \text{ ft}$$

$$H_L = S_f \times L = (4 \text{ ft}/100 \text{ ft}) \times 985 \text{ ft} = 39.4 \text{ ft}$$

$$\text{Motor hp} = \gamma \times Q \times H/550 \times ep$$

Where

g = 62.4 lb/cu ft;
Q = flowrate, cfs;
H = head, ft; and
ep = pump efficiency.

$$\frac{0.56 \text{ cfs} \times 62.4 \text{ lb/cu ft} \times 39.4 \text{ ft}}{550 \text{ ft-lb/sec/hp} \times 0.70} = 3.6 \text{ hp}$$

Note that the wire-to-water efficiency is not used here because we are only trying to find the motor output horsepower not the input electrical energy. Also, we will assume that the sludge has the same unit weight as water. The unit weight could easily be adjusted if necessary, if it were known. Just multiply the 62.4 lb/cu ft by the specific gravity of the sludge.

Comment: It might be appropriate to use a 5- or 7.5-hp motor.

Head loss for water

$$V = 1.318\ C_H\ R^{0.63}\ S_f^{0.54}$$

$$R = \text{hydraulic radius} = D/4 \text{ for pipes flowing full}$$

$$R = 0.5\text{ ft}/4 = 0.125\text{ ft};\ R^{0.63} = (0.125)^{0.63} = 0.27$$

Substituting into the Hazen-Williams equation:

$$2.8 = 1.318 \times (140) \times (0.27)\ S_f^{0.54}$$

$$S_f^{0.54} = 2.8/(1.318 \times 140 \times 0.27) = 0.056$$

$$S_f = (0.056)^{1/0.54} = (0.056)^{1.85} = 0.0049\text{ ft/ft}$$

Comment: The value for S_f should be identical in International Standard units and U.S. customary units. The difference is because of rounding of flowrates and so on.

$$H_f = S_f \times L = 0.0049\text{ ft/ft} \times 985\text{ ft} = 4.8\text{ ft}$$

Note that the method used to determine the sludge head loss is not the only method available. Several methods are discussed in various manuals and texts (Metcalf and Eddy/AECOM, 2013; WEF et al., 2018). It is recommended that the head loss be determined using various methods and then selecting the design value conservatively.

SOLUTION 8.14

280 m (440 psi)

In International Standard units:

$$Q = 125\text{ L/s} \times (1\text{ m}^3/1000\text{ L}) = 0.125\text{ m}^3/\text{s}$$

$$\text{Area} = A = (\pi D^2)/4 = [3.14(0.25\text{ m})^2]/4 = 0.049\text{ m}^2$$

$$V = Q/A = 0.125\text{ m}^3/\text{s} \div 0.049\text{ m}^2 = 2.5\text{ m/s}$$

Turning of the pump suddenly will cause a worst-case change in velocity of 2.5 m/s (full flow to complete stop).

Use $a = 1100$ m/s

$$DH = a \times DV/g = (1100 \text{ m/s}) \times (2.5 \text{ m/s})/(9.81 \text{ m/s}^2) = 280 \text{ m}$$

In U.S. customary units:

$Q = 2000$ gpm $\times$ 1 cu ft/7.48 gal $\times$ 1 min/60 sec $= 4.45$ cfs

Area $= A = (\pi D^2)/4 = [3.14(0.83 \text{ ft})^2]/4 = 0.54$ sq ft

$V = Q/A = 4.45$ ft/sec $\div$ 0.54 sq ft $= 8.2$ ft/sec

Use $a = 4000$ ft/sec

$\Delta H = a \times DV/g = (4000 \text{ ft/sec}) \times (8.2 \text{ ft/sec})/(32.2 \text{ ft/sec}^2) = 1020 \text{ ft} \times 62.4 \text{ lb/cu ft} \times$ 1 sq ft/144 sq in. $= 440$ psi

Comment: These pressure rises are very high, but represent a worst-case condition, typically associated with a power outage or inadvertent stopping of the pump. To keep these pressures from causing damage, surge control devices are typically installed or the velocity of flow should be reduced by making the pipe larger.

SOLUTION 8.15

4.5 seconds

In International Standard units:

$T_c = 2 \times L/a$

If the valve closure time is less than T_c, the valve closure can be considered instantaneous and the pressure rise determined in Problem 8.14 can be considered reasonable estimates of the pressure rise.

$L = 2500$ m

$a = 1100$ m/s

$T_c = (2 \times 2500 \text{ m})/(1100 \text{ m/s}) = 4.5$ s

In U.S. customary units:

$L = 8200$ ft

$a = 4000$ ft/sec

$T_c = 2 \times 8200 \text{ ft}/4000 \text{ ft/sec} = 4.5$ sec

Comment: When closing a valve, recommended time is $3 \times T_c$, with $2.5 \times T_c$ taken during the last portion of the closure. Recommended closure time would be 13 to 15 seconds (Jones et al., 2006).

SOLUTION 8.16

0.34 m (0.86 ft)

In International Standard units:

Because we are dealing with a chemical solution that is not water, we must use the Darcy-Weisbach equation.

$$H_f = f \times L/D \times V^2/2g$$

Where

> f = Darcy-Weisbach friction factor, which must be determined from the Moody diagram (1944) as a function of the Reynold's number and the relative roughness of the pipe ε/D;
> L = pipe length, m;
> D = pipe diameter, m; and
> V = flow velocity, m/s.

$$N_R = V \times D \times \rho/\mu$$

Where

> N_R = Reynold's number;
> ρ = density, kg/m^3;
> μ = viscosity, N·s/m^2; and
> H_f = friction loss, m/m.

Before we begin the solution, a primer on viscosity is appropriate. There are two forms of viscosity: absolute viscosity (or "dynamic viscosity" or just "viscosity") and kinematic viscosity. The way to tell them apart is by their symbol or better yet, the units. Viscosity, μ, has the units of N·s/m^2, or lb-sec/sq ft in U.S. customary units.

Kinematic viscosity, ν, has the units of m^2/s, or sq ft/sec in U.S. customary units; kinematic viscosity is related to viscosity as follows: $\nu = \mu/\rho$.

Furthermore, in International Standard units, "poise", or, more commonly, "centipoise", is frequently used for viscosity and "Stokes" or "centistokes" is frequently used for kinematic viscosity. In these cases, "centi" = 10^{22}.

$1 \text{ N·s/m}^2 = 10 \text{ poise} = 10^3 \text{ centipoise or cp [water at ambient temperature} \approx 1 \text{ cp]}$

$1 \text{ m}^2/\text{s} = 10^4$ Stokes $= 10^6$ centistokes or cs [water at ambient temperature ≈ 1 cs]

$Q = 0.06$ L/s $\times$ (1 m³/1000 L) $= 6 \times 10^{-5}$ m³/s

Area $= A = (\pi D^2)/4 = [3.14(0.018 \text{ m})^2]/4 = 2.54 \times 10^{-4}$ m²

$V = Q/A = 6 \times 10^{25}$ m³/s $\div 2.54 \times 10^{24}$ m² $= 0.24$ m/s

$\rho = 1.53 \times 1$ kg/L $\times 1000$ L/m³ $= 1.53 \times 10^3$ kg/m³

$\mu = 40$ cp $\times 1$ N·s/m² $\div 10^3$ cp $= 40 \times 10^{-3}$ N·s/m²

$N_R = V \times D \times \rho/\mu;$

$N_R = [(0.24 \text{ m/s}) \times (0.018 \text{ m}) \times (1.53 \times 10^3 \text{ kg/m}^3)]/(40 \times 10^{-3} \text{ N·s/m}^2)$

$N_R = 165$ (note that the Reynold's number is dimensionless)

Because $N_R < 1000$, this is laminar flow

For laminar flow, $f = 64/N_R = 64/165 = 0.39$

$H_f = f \times L/D \times V^2/2g = (0.39) \times (50 \text{ m}/0.018 \text{ m}) \times (0.24 \text{ m/s})^2/(2 \times 9.81 \text{ m/s}^2)$

$H_f = 3.2$ m

Compare to water:

$N_R = V \times D \times \rho/\mu;$

$\rho = 1.0$ kg/L $\times 1000$ L/m³ $= 10^3$ kg/m³

$\mu = 1$ cp $\times 1$ N·s/m² $\div 10^3$ cp $= 10^{23}$ N·s/m²

$N_R = (0.24 \text{ m/s}) \times (0.018 \text{ m}) \times (10^3 \text{ kg/m}^3)/10^{-3} \text{ N·s/m}^2$

$N_R = 4320$

Because $N_R > 1000$, it is turbulent flow

From the Moody diagram (1944) (see below):

$f = 0.042$

$H_f = f \times L/D \times V^2/2g = (0.042) \times (50 \text{ m}/0.018 \text{ m}) \times (0.24 \text{ m/s})^2/(2 \times 9.81 \text{ m/s}^2)$

$H_f = 0.34$ m (approximately 10% of the head loss of the sodium hydroxide)

In U.S. customary units:

$Q = 1$ gpm $\times$ 1 cu ft/7.48 gal $\times$ 1 min/60 sec = 0.0022 cfs

Diameter = D = 0.75 in. $\times$ 1 ft/12 in. = 0.0625 ft

Area = $A = (\pi D^2)/4 = [3.14(0.0625 \text{ ft})^2]/4 = 0.0031$ sq ft

$V = Q/A = 0.0022$ ft/s/0.0031 sq ft = 0.71 ft/sec

$m = 83.5 \times 10^{-5}$ lb-sec/sq ft

Unit weight, g, of the 50% sodium hydroxide = 1.53 $\times$ 62.4 lb/cu ft = 95.5 lb/cu ft

$\rho = \gamma/g$ = 95.5 lb/cu ft $\div$ 32.2 ft/sec^2 = 2.96 slug/cu ft

$N_R = V \times D \times \rho/m$

$N_R = (0.71 \text{ ft/sec}) \times (0.0625 \text{ ft}) \times (2.96 \text{ slug/cu ft})/(83.5 \times 10^{-5} \text{ lb-sec/sq ft})$

$N_R = 157$ (note that the Reynold's number is dimensionless); this should be identical to that for the International Standard system, but rounding introduces a small error.

Because $N_R < 1000$, this is laminar flow

For laminar flow, $f = 64/N_R = 64/157 = 0.40$

$H_f = f \times L/D \times V^2/2g = (0.40) \times (165 \text{ ft}/0.0625 \text{ ft}) \times (0.71 \text{ ft/sec})^2/(2 \times 32.2 \text{ ft/sec}^2)$

$H_f = 8.3$ ft (again differs slightly because of rounding)

Compare to water:

$Q = 1$ gpm $\times$ 1 cu ft/7.48 gal $\times$ 1 min/60 sec = 0.0022 cfs

Diameter = D = 0.75 in. $\times$ 1 ft/12 in. = 0.0625 ft

Area = $A = (\pi D^2)/4 = [3.14(0.0625 \text{ ft})^2]/4 = 0.0031$ sq ft

$V = Q/A = 0.0022$ ft/sec $\div$ 0.0031 sq ft = 0.71 ft/sec

$\mu = 2.05 \times 10^{-5}$ lb-sec/sq ft

$\rho = \gamma/g$ = 62.4 lb/cu ft/32.2 ft/sec^2 = 1.94 slug/cu ft

$N_R = V \times D \times \rho/\mu$;

$N_R = (0.71\ \text{ft/sec}) \times (0.0625\ \text{ft}) \times (1.94\ \text{slug/cu ft})/(2.05 \times 10^{-5}\ \text{lb-sec/sq ft})$

$N_R = 4200$

Because $N_R > 1000$, this is turbulent flow

$f = 0.042$ from the Moody diagram (1944)

$H_f = f \times L/D \times V^2/2g = (0.042) \times (165\ \text{ft}/0.0625\ \text{ft}) \times (0.71\ \text{ft/sec})^2/(2 \times 32.2\ \text{ft/sec}^2)$

$H_f = 0.86$ ft (again, differs slightly from the International Standard unit equivalent because of rounding)

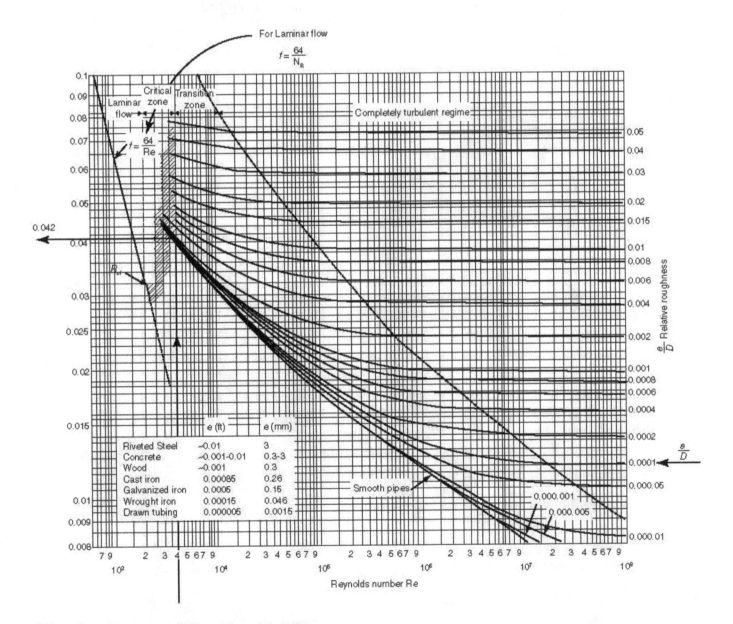

Moody diagram (Moody, 1944)

References

Jones, G. M.; Sanks, R. L.; Tchobanoglous, G.; Bossermann, B. E., II (Eds.) (2006) *Pump Station Design*, 3rd ed.; Butterworth-Heinemann: Burlington, Massachusetts.

Metcalf and Eddy, Inc./AECOM (2013) *Wastewater Engineering Treatment and Resource Recovery*, 5th ed.; McGraw-Hill: New York.

Moody, L. F. (1944) Friction Factors for Pipe Flow. *Trans. ASME*, **66**, 671–684.

Mulbarger, M. C.; Copas, S. R.; Kordic, J. R.; Cash, F. M. (1981) Pipeline Friction Losses for Wastewater Sludges. *J. Water Pollut. Control Fed.*, **53**, 1303.

Water Environment Federation; American Society of Civil Engineers; Environmental and Water Resources Institute (2018) *Design of Water Resource Recovery Facilities*, 6th ed.; WEF Manual of Practice No. 8/ASCE Manuals and Reports on Engineering Practice No. 76; Water Environment Federation: Alexandria, Virginia.

Chapter 9
CHEMICAL STORAGE, HANDLING, AND FEEDING

Problems

PROBLEM 9.1

Storage silos for dry lime are located inside a building because

    a) the lime will not bridge.
    b) it is easier to access and maintain the equipment.
    c) airtight bins are not needed to guard against moisture.

PROBLEM 9.2

The type of liquid feeder required depends on all but which of the following characteristics of the liquid chemical?

    a) Corrosivity
    b) Stability
    c) Viscosity

PROBLEM 9.3

What organization is primarily responsible for employee protection from chemicals in the workplace?

    a) U.S. Environmental Protection Agency
    b) National Fire Protection Association
    c) Occupational Safety and Health Administration

PROBLEM 9.4

Liquid sodium hypochlorite is becoming more popular than gaseous chlorine because

a) liquid sodium hypochlorite is neutral in strength and will not affect the final discharge limitations.

b) sodium hypochlorite does not lose strength with storage time.

c) ventilation gas scrubbers may be required for a catastrophic loss of gaseous chlorine.

PROBLEM 9.5

Ferric chloride should not be conveyed (piped) in

a) stainless steel.

b) high-density polyethylene pipe.

c) fiberglass-reinforced plastic pipe.

PROBLEM 9.6

An evaporator is used to

a) increase the rate of flow of liquid chlorine or sulfur dioxide to an educator.

b) create a vacuum to convey gaseous chlorine or sulfur dioxide under negative pressure for safety considerations.

c) convert liquid chlorine or sulfur dioxide to a gas.

PROBLEM 9.7

A chemical day tank is situated adjacent to a bulk storage tank. The day tank needs to be elevated above the base of the bulk storage tank to

a) allow for a gravity feed of chemical from the day tank to the point of application.

b) provide for secondary containment structure under the day tank to capture any loss of chemical.

c) prevent a potential overflow of the day tank as a result of high liquid level in the bulk storage tank.

PROBLEM 9.8

Liquid bulk storage tanks should be sized

    a)   according to the maximum daily use of the chemical.

    b)   to provide at least 60 days of chemical use under average day demand.

    c)   according to the average daily chemical use.

    d)   to accommodate twice the maximum delivery volume capacity of the hauler.

PROBLEM 9.9

Glass-lined piping is not acceptable for

    a)   anionic polymers.

    b)   ferrous sulfate.

    c)   hydrofluoric acid.

PROBLEM 9.10

Double-wall chemical storage tanks do not need to be placed within a secondary containment structure.

    a)   True

    b)   False

PROBLEM 9.11

Federal regulations require that material safety data sheets for on-site chemicals be made available to employees at the treatment works.

    a)   True

    b)   False

PROBLEM 9.12

Liquid ferric chloride will not freeze; therefore, it can be stored in bulk outdoors.

    a)   True

    b)   False

PROBLEM 9.13

Sodium hypochlorite can be generated on demand through the electrolysis of a _____ solution.

PROBLEM 9.14

In a chlorine storage room, ventilation needs to be drawn from the floor because chlorine gas is _____ than air.

PROBLEM 9.15

The concentration of phosphorus in a wastewater flow of 567 m³/d (0.15 mgd) averages 6.0 mg/L. Ferric chloride, 37%, specific gravity of 1.386 is to be used to precipitate the phosphorus. The state requires the addition of five units of ferric chloride per unit of phosphorus to compensate for influent variations. The average daily ferric chloride use requirement is

    a)   1.26 L/d (0.33 gpd)
    b)   6.67 L/d (1.76 gpd)
    c)   12.3 L/d (3.25 gpd)
    d)   33.1 L/d (8.72 gpd)

PROBLEM 9.16

Using data from Problem 9.15, it is desired to add the ferric chloride to the raw wastewater flow in the pipe manifold immediately following influent pumping to provide good mixing of the chemical. The influent pumping rate is 2720 m³/d (500 gpm) at constant speed. What is the required chemical feed rate to be paced with pumping?

    a)   0.23 L/min (0.061 gpm)
    b)   0.11 L/min (0.029 gpm)
    c)   0.09 L/min (0.024 gpm)
    d)   0.04 L/min (0.011 gpm)

PROBLEM 9.17

A containment wall structure is to be built around two 31-m³ (8000-gal) storage tanks for liquid alum. Each tank has a base diameter of 3.0 m (9.84 ft). A containment area (inside dimensions) of 6 m by 10 m (19.7 ft × 32.81 ft) is to be used for the containment structure. The state regulation is to provide containment for 150% of the contents of the largest tank in the containment area. What is the minimum height of the containment wall?

    a)   0.78 m (2.56 ft)

    b)   0.88 m (2.81 ft)

    c)   1.03 m (3.38 ft)

    d)   1.75 m (5.74 ft)

Solutions

SOLUTION 9.1

    b)   it is easier to access and maintain the equipment.

Silos for dry lime located inside a building can be made for easy access and maintenance and are not hindered by bad or cold weather.

SOLUTION 9.2

    b)   Stability

Stability does not affect metering pump selection. If a chemical has lost its stability, it is probably too old and should not be used.

SOLUTION 9.3

    c)   Occupational Safety and Health Administration

The Occupational Safety and Health Administration's Process and Safety Management standard is intended for employee protection from chemicals in the workplace.

SOLUTION 9.4

    c)   ventilation gas scrubbers may be required for a catastrophic loss of gaseous chlorine.

SOLUTION 9.5

    a)   stainless steel.

The chlorides will attack stainless steel and promote corrosion.

SOLUTION 9.6

c)   convert liquid chlorine or sulfur dioxide to a gas.

If chlorine or sulfur dioxide is drawn as a liquid, an evaporator is used to vaporize the gas.

SOLUTION 9.7

c)   prevent a potential overflow of the day tank as a result of high liquid level in the bulk storage tank.

The maximum elevation of chemical in a day tank needs to be higher than the maximum elevation of chemical in an associated bulk storage tank. If the day tank is at a lower elevation, the head of liquid in the storage tank could cause an overflow of chemical in the day tank. The layout should allow the contents of the storage tank to be pumped to the day tank.

SOLUTION 9.8

c)   according to the average daily chemical use.

SOLUTION 9.9

c)   hydrofluoric acid.

Hydrofluoric acid will rapidly attack silicates in the glass.

SOLUTION 9.10

b)   False

Many regulatory agencies still require secondary containment. If there is a catastrophic failure of the tank, then all contents would be lost if not contained.

SOLUTION 9.11

a)   True

SOLUTION 9.12

b)   False

The chemical has a low freezing point depending on concentration, but it can freeze.

SOLUTION 9.13

Sodium hypochlorite can be generated on demand through the electrolysis of a <u>brine</u> solution.

Salt (sodium chloride) is dissolved in water in a tank and through electrolysis the chlorides are converted to chlorine and dissolved in water to form a hypochlorite solution.

SOLUTION 9.14

In a chlorine storage room, ventilation needs to be drawn from the floor because chlorine gas is <u>heavier</u> than air.

If it escapes from the container, it will settle to the floor. Consequently, ventilation ducts should remove the gas from near the floor.

SOLUTION 9.15

d)   33.1 L/d (8.72 gpd)

In International Standard units:

$$6 \text{ mg/L} \times 567 \text{ m}^3/\text{d}\left(\frac{0.001 \text{ kg/m}^3}{\text{mg/L}}\right) = 3.4 \text{ kg/d phosphorus}$$

Unit weight of 37% ferric chloride is = 1.386 × 1.0 kg/L = 1.386 kg/L aqueous solution

For a 37% solution, 0.37 × 1.386 kg ferric chloride/L = 0.513 kg/L ferric chloride

Daily chemical required:

$$\frac{3.4 \text{ kg/d phosphorus} \times 5 \text{ kg ferric chloride/kg phosphorus}}{0.513 \text{ kg/L ferric chloride}} = 33.1 \text{ L/d ferric chloride (37\% solution)}$$

In U.S. customary units:

$$6 \text{ mg/L} \times 0.15 \text{ mgd}\left(\frac{8.34 \text{ lb/mil. gal}}{\text{mg/L}}\right) = 7.5 \text{ lb/d phosphorus}$$

Unit weight of 37% ferric chloride is 1.386 × 8.34 lb/gal = 11.6 lb/gal aqueous solution

For a 37% solution, 0.37 × 11.6 lb ferric chloride/gal = 4.3 lb/gal ferric chloride

Daily chemical required:

$$\frac{7.5 \text{ lb/d phosphorus} \times 5 \text{ lb ferric chloride/lb phosphorus}}{4.3 \text{ lb ferric chloride/gal}} = 8.72 \text{ gpd ferric chloride (37\% solution)}$$

SOLUTION 9.16

b)   0.11 L/min (0.029 gpm)

In International Standard units:

$$6 \text{ mg/L} \times 2720 \text{ m}^3/\text{d} \times \left(\frac{0.001 \text{ kg/m}^3}{\text{mg/L}}\right) = 16.32 \text{ kg/d phosphorus pumped}$$

$$\frac{16.32 \text{ kg/d}}{24 \text{ h/d} \times 60 \text{ min/h}} = 0.011\,33 \text{ kg/min phosphorus}$$

$$\frac{(0.011\,33 \text{ kg/min}) \times (5 \text{ kg ferric chloride/kg phosphorus})}{0.513 \text{ kg ferric chloride/L}} = 0.11 \text{ L/min ferric chloride (37\% solution)}$$

In U.S. customary units:

$$500 \text{ gpm} \times 1440 \text{ min/d} = 720\,000 \text{ gpd or } 0.72 \text{ mgd}$$

$$6 \text{ mg/L} \times 0.72 \text{ mgd} \times \left(\frac{8.34 \text{ lb/mil. gal}}{\text{mg/L}}\right) = 36 \text{ lb/d phosphorus pumped}$$

$$\frac{36 \text{ lb/d}}{1440 \text{ min/d}} = 0.025 \text{ lb/min phosphorus}$$

$$\frac{(0.025 \text{ lb/min}) \times (5 \text{ lb ferric/lb phosphorus})}{4.3 \text{ lb/gal}} = 0.029 \text{ gpm ferric chloride (37\% solution)}$$

SOLUTION 9.17

b)   0.88 m (2.81 ft)

In International Standard units:

Area of base of one tank $= A = 3.14(1.5 \text{ m})^2 = 7.1 \text{ m}^2$

Volume to contain: $1.5 \times 31 \text{ m}^3 = 46.5 \text{ m}^3$

Inside area for containment: $6 \text{ m} \times 10 \text{ m} = 60 \text{ m}^2$

Delete area of one tank. It is assumed that only one tank will fail. Therefore, the area occupied by the second tank is not available for containment.

Net area: $60 \text{ m}^2 - 7.1 \text{ m}^2 = 52.9 \text{ m}^2$

Height of containment wall:

$46.5 \text{ m}^3/52.9 \text{ m}^2 = 0.88 \text{ m}$

In U.S. customary units:

Area of base of one tank $= A = 3.14(4.92 \text{ ft})^2 = 76 \text{ sq ft}$

$$\text{Volume to contain: } 1.5 \times 8000 \text{ gal} = \frac{12\ 000 \text{ gal}}{7.48 \text{ gal/cu ft}} = 1604 \text{ cu ft}$$

Inside area for containment: $19.69 \text{ ft} \times 32.81 \text{ ft} = 646 \text{ sq ft}$

Delete area of one tank. It is assumed that only one tank will fail. Therefore, the area occupied by the second tank is not available for containment.

Net area: $646 \text{ sq ft} - 76 \text{ sq ft} = 570 \text{ sq ft}$

Height of containment wall:

$1604 \text{ cu ft}/570 \text{ sq ft} = 2.81 \text{ ft}$

Chapter 10
ELECTRICAL DISTRIBUTION SYSTEMS

Problems

PROBLEM 10.1

Voltage in water resource recovery facilities (WRRFs) is typically higher than the 120 V typically used in homes because

   a) higher voltages are safer for the electrical maintenance staff.
   b) higher voltages can power higher efficiency motors.
   c) higher voltages reduce the power factor.

PROBLEM 10.2

What is the purpose of a single line distribution diagram?

   a) It allows the power company to distribute the power more efficiently.
   b) It shows the power sources for all units drawing power and all feeders consuming power.
   c) It shows the emergency generator capacity and transfer switches.

PROBLEM 10.3

Why is standby power needed at a WRRF and how is it typically supplied?

PROBLEM 10.4

Three ways in which electric utilities charge their clients for electricity use are _____, _____, and _____.

PROBLEM 10.5

Types of adjustable-speed drives include

    a)   variable-frequency drives, belt drives, and eddy current drives.
    b)   squirrel cage drives and pneumatic system drives.
    c)   adjustable-voltage drives, gear drives, and hydraulic current drives.

PROBLEM 10.6

When the current and the voltage are not exactly in phase, what results? How are these results counteracted?

PROBLEM 10.7

What is the importance of a relay coordination study?

    a)   Breakers in the study are interchangeable in case of a failure.
    b)   Breakers are sized so that a power failure involves the largest area.
    c)   Breakers are sized so that a power failure is isolated to the smallest area.

PROBLEM 10.8

What are minimum safety precautions for working with energized electrical equipment?

PROBLEM 10.9

Name the recommended steps for regular preventive maintenance on low-voltage motor control centers?

    a)   Test insulation resistance, clean and inspect, check for signs of overheating, verify that indicator lights are working
    b)   No preventive maintenance is necessary because it is low voltage
    c)   Test voltage regulator, clean heaters, check ground wires

PROBLEM 10.10

What factors should be considered when evaluating the cost-effectiveness of co-generation?

PROBLEM 10.11

Which of the following changes can be implemented to improve the energy efficiency of a facility?

a) Replace fluorescent lighting with incandescent lighting.
b) Install energy-efficient windows, lighting, and equipment.
c) Bring power in at 120 V rather than 480 V.

PROBLEM 10.12

In a typical facility, the electrical distribution system is a wye system that economically provides what voltage for lighting?

a) 120
b) 408
c) 277
d) 420

Solutions

SOLUTION 10.1

b) Higher voltages can power higher efficiency motors.

SOLUTION 10.2

b) It shows the power sources for all units drawing power and all feeders consuming power.

SOLUTION 10.3

Standby power is needed to ensure that wastewater treatment processes remain functional during loss of primary power. Standby power is typically supplied through a generator or a secondary or backup feed or both. Switching of power supply to the generator is typically accomplished via an automatic transfer switch, whereas switchover to a secondary feed is typically accomplished by manual transfer.

SOLUTION 10.4

Three ways in which electric utilities charge their clients for electricity include any three of the following:

- Flat fee,
- Usage rate ($/kWH),
- Demand charges (kW),
- Power factor charges, and
- Off-peak/peak rate adjustments.

SOLUTION 10.5

a) variable-frequency drives, belt drives, and eddy current drives.

SOLUTION 10.6

When the current and voltage are out of phase, the power factor is decreased. In the case of a switch mode power supply, harmonics are created. The power factor can be returned to near unity by adding capacitors (or inductors) or an unloaded synchronous motor to the system.

SOLUTION 10.7

c) Breakers are sized so that a power failure is isolated to the smallest area.

A relay coordination study ensures that breakers in the system are properly sized so that a power failure is isolated in the most downstream (smallest) area possible. If breakers are not sized and located properly, a failure in a small branch line may affect the entire distribution system.

SOLUTION 10.8

Work on energized electrical equipment should be avoided whenever possible. However, if work on an energized system is necessary, the following precautions should be followed:

- Ensure that work is being performed by authorized and experienced personnel,
- Use ample lighting to make repairs,
- Insulate employees from electrical sources using dry wood or a rubber mat,
- Use one hand to make repairs when possible,

- Identify all circuit breakers and establish plan for de-energizing equipment in the case of emergency, and
- Have a person trained in first aid for electrical shock near the work area.

SOLUTION 10.9

a) Test insulation resistance, clean and inspect, check for signs of overheating, verify that indicator lights are working

SOLUTION 10.10

Co-generation is applicable to facilities using anaerobic digestion for solids processing. Factors that influence the cost-effectiveness of these types of systems include the following:

- Cost of power/natural gas—the higher the cost of power or natural gas from the local utility, the more economically attractive co-generation becomes.
- Cost of equipment for co-generation—the capital cost for any equipment used for co-generation must be figured into the economic analysis. These costs are typically offset by the power or natural gas savings associated with the system. Equipment included for co-generation includes
  - Co-generation device;
  - Gas storage, handling, and preparation equipment; and
  - Electrical gear for parallel supply.
- Potential credit from utility—in some cases, a utility may buy back power or give a credit to facilities that can have interruptible power supply (i.e., switch to generators when grid power is low). Co-generation facilities would provide excess power for utility buyback and will influence the ability of the facility to be able to use alternative power supply.

SOLUTION 10.11

b) Install energy-efficient windows, lighting, and equipment.

SOLUTION 10.12

c) 277

Chapter 11
UTILITIES

Problems

PROBLEM 11.1

A facility has a wet well at a pumping station that is 3.7 m (12 ft) in diameter and 4.7-m (15-ft) deep. Assuming that the wet well has an average water level of 0.9 m (3 ft), what is the recommended minimum airflow in cubic meters per minute (cubic feet per minute) to continuously ventilate the wet well?

PROBLEM 11.2

The air compressor for a dewatering facility can produce a maximum of 8.5 m^3/min (300 cfm). If a heatless-type air dryer is added to the unit to improve air quality, what is the minimum airflow the compressor can produce after drying?

PROBLEM 11.3

A facility needs to decant its aerobic digester using a portable trash pump. The digester is 21.3-m (70-ft) square, with a sidewater depth of 3.7 m (12 ft). Assuming that the maximum pumping rate is 18.9 L/s (300 gpm), what is the minimum required operation time (to the closest half hour) to draw the digester down to a sidewater depth of 1.2 m (4 ft)?

PROBLEM 11.4

Water resource recovery facilities should be protected against a maximum of a 25-year flood.

  a) True
  b) False

PROBLEM 11.5

What should hose outlets be equipped with to prevent contamination of the drinking water supply?

a) Antisiphon devices
b) Cross connections
c) Pressure-relief valves

PROBLEM 11.6

What are the three most common types of heating for buildings or process equipment areas?

a) Infrared, steam, and hot air
b) Hot water, infrared, and steam
c) Hot water, steam, and hot air

PROBLEM 11.7

Solid waste shall be stored in a suitable container to prevent the risk of fire and _____?

a) Stormwater pollution
b) Vermin infestation
c) Grease accumulation

PROBLEM 11.8

List the proper procedures in the event that the odor of natural gas is detected.

PROBLEM 11.9

If a fire begins within the motor control center and electrical distribution gear, what is the appropriate designation of fire extinguisher to be used?

a) B
b) C
c) D

PROBLEM 11.10

What is the minimum recommended interval for inspection of fan wheels, shafts, and housing scrolls for air handling units?

a) Monthly
b) Annually
c) Semiannually

PROBLEM 11.11

The rotary vane, rotary screw, and diaphragm are all examples of what type of compressor?

a) Dynamic
b) Rotary
c) Positive displacement

PROBLEM 11.12

Chlorides from deicing of walkways and roadways can degrade concrete surfaces.

a) True
b) False

Solutions

SOLUTION 11.1

8.2 m³/min (271 cfm)

The recommended minimum air change per hour is 12. The volume of air space in the wet well under average conditions is calculated as follows:

In International Standard units:

Surface area of wet well $(3.14/4) \times (3.7 \text{ m})^2 = 10.75 \text{ m}^2$

Volume $10.75 \text{ m}^2 \times (4.7 \text{ m} - 0.9 \text{ m}) = 41 \text{ m}^3$

In U.S. customary units:

$$\text{Surface area of wet well } (3.14/4) \times (12 \text{ sq ft}) = 113 \text{ sq ft}$$

$$\text{Volume} \quad 113 \text{ sq ft} \times (15 \text{ ft} - 3 \text{ ft}) = 1356 \text{ cu ft}$$

At 12 air changes/h, the minimum airflow rate is as follows:

In International Standard units:

$$(12 \text{ air changes/h}) \times (41 \text{ m}^3/\text{air changes}) \times (h/60 \text{ min})$$

$$= 8.2 \text{ m}^3/\text{min}$$

In U.S. customary units:

$$(12 \text{ air changes/h}) \times (1359 \text{ cu ft}/\text{air changes}) \times (h/60 \text{ min})$$

$$= 271 \text{ cfm}$$

SOLUTION 11.2

7.2 m³/min (255 cfm)

Heatless air dryers consume approximately 10 to 15% of the airflow to purge the adsorption media. To find the minimum airflow, reduce the rate by the maximum consumption (15%).

In International Standard units:

$$(8.5 \text{ m}^3/\text{min}) - (8.5 \text{ m}^3/\text{min} \times 0.15) = 7.2 \text{ m}^3/\text{min}$$

In U.S. customary units:

$$300 \text{ cfm} - (300 \text{ cfm} \times 0.15) = 255 \text{ cfm}$$

SOLUTION 11.3

Approximately 16.5 hours

The required volume to be removed from the digester is as follows:

In International Standard units:

$$(21.3 \text{ m} \times 21.3 \text{ m}) \times (3.7 \text{ m} - 1.2 \text{ m}) = 1134 \text{ m}^3$$

$$1134 \text{ m}^3 \times 1000 \text{ L/m}^3 = 1\,134\,000 \text{ L}$$

In U.S. customary units:

$$(70 \text{ ft} \times 70 \text{ ft}) \times (12 \text{ ft} - 4 \text{ ft}) = 39\,200 \text{ cu ft}$$

$$39\,200 \text{ cu ft} \times 7.48 \text{ gal/cu ft} = 293\,216 \text{ gal}$$

At a pumping rate of 18.9 L/s (300 gpm), the time required is as follows:

In International Standard units:

$$1\,134\,000 \text{ L}/(18.9 \text{ L/s}) = 60\,000 \text{ s} = 16 \text{ h } 40 \text{ min}$$

In U.S. customary units:

$$293\,216 \text{ gal}/300 \text{ gpm} = 977 \text{ min} = 16 \text{ h } 17 \text{ min}$$

SOLUTION 11.4

b)   False

Facilities should be fully protected against a maximum of a 100-year flood.

SOLUTION 11.5

b)   Antisiphon devices

SOLUTION 11.6

c)   Hot water, steam, and hot air

SOLUTION 11.7

b) Vermin infestation

SOLUTION 11.8

Turn off electrical power to avoid sources of ignition. Determine the source of the natural gas and, if gas smell is severe, evacuate the area and immediately notify the local utility and fire department.

SOLUTION 11.9

b) C

SOLUTION 11.10

b) Annually

SOLUTION 11.11

c) Positive displacement

SOLUTION 11.12

a) True

Chapter 12
MAINTENANCE

Problems

PROBLEM 12.1

List the four categories into which maintenance strategies fall.

PROBLEM 12.2

List the three roots of failure.

PROBLEM 12.3

Identify five of the seven pump installation fundamentals.

PROBLEM 12.4

Couplings do not correct, absorb, dampen, or otherwise fix shaft misalignment in any way.

   a) True
   b) False

PROBLEM 12.5

The resulting force caused by the unbalance does not increase with speed, making proper balancing critical at all operating speeds.

   a) True
   b) False

PROBLEM 12.6

Piping strain in excessive amounts can make a precision balance and shaft alignment job pointless.

a) True
b) False

PROBLEM 12.7

How do you check for piping strain and what is the maximum movement you should see?

PROBLEM 12.8

What is soft foot condition?

PROBLEM 12.9

When using shims for alignment, be sure to use shims that provide

a) full footprint support across the face.
b) a skidproof surface.
c) equal thickness.

PROBLEM 12.10

A torque wrench is needed for good installation practices.

a) True
b) False

PROBLEM 12.11

What does vibration analysis not identify?

a) Machine looseness
b) Misalignment
c) Lack of lubrication

PROBLEM 12.12

What does an accelerometer do?

PROBLEM 12.13

Lubrication and wear particle analysis provide information on the condition of the lubricant itself as well as the wear condition of the friction surfaces the lubricant is protecting.

a)   True
b)   False

PROBLEM 12.14

List three of the four conditions tracked by analysis of the lubricant.

PROBLEM 12.15

List three of the four wear characteristics gathered by wear particle analysis.

PROBLEM 12.16

Identify five types of equipment that are typically tested using thermography power?

PROBLEM 12.17

Motor life is reduced by 50% for every 10 °C rise above its rating.

a)   True
b)   False

PROBLEM 12.18

Ultrasonic sound waves have frequencies above the threshold at which human hearing stops.

a)   20 kHz
b)   20 dB
c)   20 Hz

PROBLEM 12.19

A useful application of ultrasonic technology is determining when to grease bearings and how much grease to apply.

a) True
b) False

PROBLEM 12.20

A surge test is the only electrical condition monitoring that provides information on the health of electrical windings.

a) True
b) False

PROBLEM 12.21

It is possible for a motor to continue operating even though it is failing a surge test.

a) True
b) False

PROBLEM 12.22

Motor current signature analysis is useful in detecting mechanical and electrical problems in rotating equipment.

a) True
b) False

PROBLEM 12.23

List the three rules for evaluating maintenance decisions according to the reliability centered maintenance strategy.

PROBLEM 12.24

Benchmarking is the practice of measuring performance against a standard. What are the three benefits of objectively demonstrating how a facility is being operated?

PROBLEM 12.25

What is the purpose of a computerized maintenance management system?

Solutions

SOLUTION 12.1

Maintenance strategies fall into the following four categories:

- Run-to-failure,
- Preventive (time-based),
- Condition-based (can accept predictive), and
- Proactive.

SOLUTION 12.2

The following are the three roots of failure:

- Physical,
- Human, or
- Management system (latent).

SOLUTION 12.3

Five of the seven pump installation fundamentals include any of the following:

- Sized for operation,
- Dynamically balanced,
- Mounted on stable foundations,
- Coupled using precision shaft alignment tolerances,
- Checked for pipe strain, and
- Received proper lubrication.

SOLUTION 12.4

a) True

SOLUTION 12.5

b) False

SOLUTION 12.6

a) True

SOLUTION 12.7

Place dial indicators on both shafts of the piped system and loosen the foot bolts. No more than 2 mil (0.002 in.) movement should result.

SOLUTION 12.8

When the mounting feet of a machine and the mating baseplate do not make full surface contact.

SOLUTION 12.9

a) full footprint support across the face.

SOLUTION 12.10

a) True

SOLUTION 12.11

c) Lack of lubrication

SOLUTION 12.12

It measures acceleration directly and is equipped with electronic integrators to give velocity and displacement.

SOLUTION 12.13

a)   True

SOLUTION 12.14

The following are four conditions tracked by analysis of the lubricant:

- Chemical contamination,
- Molecular condition,
- Dissolved elements, and
- State of additives.

SOLUTION 12.15

The following are four wear characteristics gathered by wear particle analysis:

- Amount,
- Makeup,
- Shape, and
- Size.

SOLUTION 12.16

The following are five types of equipment that are typically tested using thermography power:

- Switchgear,
- Connections,
- Distribution lines,
- Transformers motors,
- Generators,
- Buswork,
- Bearings couplings,
- Piston rings,
- Brake drums, and
- Heat exchangers.

SOLUTION 12.17

a) True

SOLUTION 12.18

a) 20 kHz

SOLUTION 12.19

a) True

SOLUTION 12.20

b) False

This is the only test capable of identifying turn insulation deterioration.

SOLUTION 12.21

a) True

SOLUTION 12.22

a) True

SOLUTION 12.23

The following are three rules for evaluating maintenance decisions according to the reliability centered maintenance strategy:

- Identifying a failure mode,
- Applicability, and
- Effectiveness.

SOLUTION 12.24

The following are the three benefits of objectively demonstrating how a facility is being operated:

- Identifying efficiencies,
- Quantifying progress, and
- Identifying what areas need improvement.

SOLUTION 12.25

It allows the user to track equipment and the work orders associated with it.

Chapter 13
ODOR CONTROL

Problems

PROBLEM 13.1

Estimate the useful life of a carbon bed air filter removing hydrogen sulfide (H_2S) from a foul airflow, assuming the characteristics from the table below.

Airflow at 21.2 °C (70 °F)	236 L/s (500 cfm)
H_2S concentration	5 ppm
Carbon in filter	454 kg (1000 lb)
Molecular weight of H_2S	15.4 kg/mol (34 lb/mol)
Volume of 1 kg·mol of any gas at 21.2 °C (1 lb = mol of any gas at 70 °F)	10 960 L (387 cu ft)
The carbon used can adsorb 30% of its weight in H_2S	

a) 31.6 months
b) 12.0 months
c) 11.3 months

PROBLEM 13.2

Match the odorous compound with its odor description:

a) Methyl amine   1) Rotten cabbage
b) Hydrogen sulfide   2) Vinegar
c) Thiocresol   3) Fecal material
d) Skatole   4) Fruit
e) Methyl mercaptan   5) Skunk
f) Acetic acid   6) Fish
g) Acetaldehyde   7) Rotten eggs

PROBLEM 13.3

The primary reason for concern about hydrogen sulfide in a treatment system is because of its odor potential.

a) True

b) False

PROBLEM 13.4

At what approximate concentration is hydrogen sulfide first detectable to the average human nose?

a) 0.5 ppb

b) 5 ppb

c) 500 ppb

PROBLEM 13.5

Raising the pH of a waste stream will inhibit the release of compounds such as _____, but may increase the release of compounds such as _____.

PROBLEM 13.6

If you have a process or facility such as a pumping station or manhole that is releasing odors, including hydrogen sulfide, the most appropriate long-term solution is simply to enclose the area to contain the odors.

a) True

b) False

PROBLEM 13.7

You have an odorous waste stream for which you have decided that the best approach is to chemically treat it to control odors. Based on the following information, calculate the daily dosage for the listed chemicals.

Average hydrogen sulfide concentration	5 mg/L
Average flowrate	0.2185 m³/s (5 mgd)
Ferric chloride dosage	3.5 parts iron/part sulfide
Chlorine dosage (Cl_2)	10 parts chlorine/part sulfide

a)   62.3 kg/d Fe and 178 kg/d Cl_2 (137 lb/d Fe and 392 lb/d Cl_2)
b)   156 kg/d Fe and 445 kg/d Cl_2 (343 lb/d Fe and 980 lb/d Cl_2)
c)   311.5 kg/d Fe and 890 kg/d Cl_2 (686 lb/d Fe and 1960 lb/d Cl_2)
d)   623 kg/d Fe and 1780 kg/d Cl_2 (1372 lb/d Fe and 3920 lb/d Cl_2)

PROBLEM 13.8

It is possible to remove all odorous compounds from an air stream with a single wet scrubber if it is designed correctly.

a)   True
b)   False

PROBLEM 13.9

You have determined that an odorous air stream has significant concentrations of hydrogen sulfide, methyl amine, and butyraldehyde. To achieve the most effective removal of these constituents, which of the following treatment technologies would be appropriate?

a)   An activated carbon scrubber
b)   A biotrickling filter followed by either a chemical or activated carbon polishing
c)   A wet scrubber with caustic
d)   A wet scrubber with caustic/hypochlorite
e)   A multistage biotrickling filter
f)   b) or d)

PROBLEM 13.10

You have an odorous air stream that you have decided to treat using a carbon scrubber. The characteristics of this air stream are listed below. Calculate the surface area of carbon needed.

Airflow per unit	472 L/s (1000 cfm)
Number of units on line	2–3
Hydrogen sulfide concentration	10–50 ppmv

a)   5.5 m² (59 sq ft)
b)   3.73 m² (40 sq ft)
c)   3 m² (32 sq ft)

PROBLEM 13.11

You are in the process of filling out an odor complaint called into your facility. List five pieces of information that you should record on the complaint form and explain how that information would/could be used.

PROBLEM 13.12

You are operating the primary clarifiers to thicken the primary sludge being fed to the digesters. You have begun to observe an increase in odors occurring at the primary effluent overflow weirs. Suggest some immediate steps that can be taken to help reduce these odors based on the following operating conditions:

Number of units in service	8
Average hydraulic detention time	5 h
Influent dissolved oxygen concentration	0 mg/L
Average sludge retention time	4 h
Average sludge concentration	6.5% total solids
Influent temperature	20 °C (70 °F)
Physical observations	Mats of sludge floating to surface and rising bubbles

a)   Add more primary units
b)   Reduce the number of primary units
c)   Increase sludge pumping rate
d)   Ignore the problem; it will go away on its own
e)   Add chemicals such as chlorine or ferric chloride to the influent
f)   None of the above
g)   Options b), c), and e) should be considered

Solutions

SOLUTION 13.1

a)   31.6 months

In International Standard units:

Calculate the carbon bed's capacity to adsorb H_2S:

Kilograms of carbon × Amount of H$_2$S carbon can adsorb

454 kg carbon × (0.30 kg of H$_2$S/kg carbon) = 136.2 kg H$_2$S

Calculate the kilograms of H$_2$S coming into the carbon bed

Concentration of H$_2$S entering filter × Airflow/Volume of 1 mol H$_2$S × Weight of 1 mol
H$_2$S = H$_2$S entering filter/min

(5 parts/1 000 000 parts) × (236 L/s) × (1 kg·mol/10 960 L) × (15.4 kg H$_2$S/kg·mol) =
0.000 001 6 kg/s H$_2$S entering the filter

Estimate carbon life:

$$\frac{\text{Kilograms H}_2\text{S the filter can remove}}{\text{Kilograms H}_2\text{S entering the filter per second}} = \text{Seconds the filter can absorb H}_2\text{S}$$

136.2 kg H$_2$S/(0.000 001 6 kg/s) = 85 125 000 s

Convert minutes to months:

85 125 000 s × (1 min/60 s) × (1 h/60 min) × (1 d/24 h) × (1 mo/30 d) = 31.6 mo

In U.S. customary units:

Calculate the carbon bed's capacity to adsorb H$_2$S:

Pounds of carbon × Amount of H$_2$S carbon can adsorb = Pounds of H$_2$S
the carbon can adsorb

1000 lb carbon × 0.30 lb H$_2$S/lb carbon = 300 lb H$_2$S

Calculate the pounds of H$_2$S coming into the carbon bed:

Concentration of H$_2$S entering filter × Airflow/Volume of 1 mol H$_2$S × Weight of
1 mol H$_2$S = H$_2$S entering filter/min

(5 parts/1 000 000 parts) × (500 cfm) × (1 lb-mol/387 cu ft) (34 lb H$_2$S/lb-mol) =
0.000 22 lb/min H$_2$S entering the filter

Estimate carbon life:

$$\frac{\text{Pounds of } H_2S \text{ the filter can remove}}{\text{Pounds } H_2S \text{ entering the filter per minute}} = \text{Minutes the filter can absorb } H_2S$$

$$300 \text{ lb } H_2S/(0.000\ 22 \text{ lb/min}) = 1\ 363\ 636 \text{ minutes}$$

Convert minutes to months:

$$1\ 363\ 636 \text{ minutes} \times (1 \text{ h}/60 \text{ min}) \times (1 \text{ d}/24 \text{ h}) \times (1 \text{ mo}/30 \text{ d}) = 31.6 \text{ mo}$$

SOLUTION 13.2

a) Methyl amine smells like (6) Fish

b) Hydrogen sulfide smells like (7) Rotten eggs

c) Thiocresol smells like (5) Skunk

d) Scatole smells like (3) Fecal material

e) Methyl mercaptain smells like (1) Rotten cabbage

f) Acetic acid smells like (2) Vinegar

g) Acetaldehyde smells like (4) Fruit

SOLUTION 13.3

b) False

While the odor potential associated with hydrogen sulfide is a significant concern in treatment systems, system personnel also need to be concerned because it can result in significant levels of corrosion, especially in confined spaces such as conveyance lines and covered tanks. Hydrogen sulfide is toxic at a concentration of approximately 100 ppm, which can occur in enclosed areas within the system. In addition, the Occupational Safety and Health Administration limits for exposure to hydrogen sulfide are 10 ppm (time-weighted average) and 15 ppm (short-term exposure).

SOLUTION 13.4

a) 0.5 ppb

The threshold for detecting the presence of hydrogen sulfide is 0.5 ppb or 0.0005 ppm. The concentration needs to be approximately 10 times higher for it to be identifiable as hydrogen sulfide.

SOLUTION 13.5

Raising the pH of a waste stream will inhibit the release of compounds such as <u>hydrogen sulfide,</u> but may increase the release of compounds such as <u>ammonia and amines</u>.

SOLUTION 13.6

b)   False

While containment may provide a short-term solution, it does not resolve the underlying issue. Containing the odorous materials could create a hazardous condition for workers having to enter the space and will likely result in significant levels of corrosion within the containment area unless the odorous gases are evacuated and treated.

SOLUTION 13.7

c)   311.5 kg/d Fe and 890 kg/d Cl_2 (686 lb/d Fe and 1960 lb/d Cl_2)

In International Standard units:

Calculate the mass of sulfide to be treated:

Flow (m³/s) × 86 400 sec/d × Concentration (1 part/1 million parts = 1000 kg/m³) × (Mass of sulfur/Mass of hydrogen sulfide) = Kilograms per day of sulfide

0.2185 m³/s × 86 400 sec/d × (5 kg sulfide/1 000 000 kg water × 1000 kg/m³) × 32.064 parts sulfur/34.08 parts hydrogen sulfide = 89 kg/d

Calculate the mass of chemical needed to react with the sulfide:

Mass of sulfide × Mass of chemical per unit sulfide = Mass of chemical needed

89 kg/d sulfide × 3.5 parts iron/part sulfide = 311.5 kg/d iron

89 kg/d sulfide × 10 parts chlorine per part sulfide = 890 kg/d chlorine

In U.S. customary units:

$$\text{Flow (mgd)} \times \text{Concentration (1 part/1 million parts)} \times \text{(8.34 lb/gal)} \times$$
$$\text{(Mass of sulfur/Mass of hydrogen sulfide)} = \text{Pounds per day sulfide}$$

$$\text{(5 mgd)} \times \text{(5 lb hydrogen sulfide/1 000 000 lb water)} \times \text{(8.34 lb/gal)} \times$$
$$\text{(32.064 parts sulfur/34.08 parts hydrogen sulfide)} = 196 \text{ lb/d sulfide}$$

$$196 \text{ lb/d sulfide} \times 3.5 \text{ parts iron/part sulfide} = 686 \text{ lb/d iron}$$

$$196 \text{ lb/d sulfide} \times 10 \text{ parts chlorine/part sulfide} = 1960 \text{ lb/d chlorine}$$

Note that the actual amount of chemical required will be greater, depending both on the amount of active ingredient in the product and on the actual reaction ratio of the chemical with hydrogen sulfide within your system.

SOLUTION 13.8

b) False

Trying to remove ammonia, hydrogen sulfide, and buteraldehyde in a single stage is problematic. Because of the wide variety of odorous compounds and their widely varying chemical characteristics, it is often necessary to install multistage wet scrubbers with treatment chemistries selected based on the constituents that are causing odors. Wet scrubbers using chemicals (hypochlorite, in particular) will impart an odor to the air stream even if odor compounds are removed. This is sometimes called the "swimming pool" odor that is commonly associated with wet scrubbers using hypochlorite.

SOLUTION 13.9

f) b) or d)

Either b) A biotrickling filter followed by either a chemical or activated carbon polishing or d) A wet scrubber with caustic/hypochlorite would be appropriate. The level and complexity of treatment and the selection of treatment technology(ies) will depend on the characteristics and complexity of the odors being treated. In all cases, the least complicated approach should be used. Biotrickling filters (or biofilters) have the advantage of being more sustainable technologies than physical/chemical treatment options.

SOLUTION 13.10

b)   3.73 m² (40 sq ft)

The face velocity for a carbon bed should be in the range of 203 to 380 mm/s (40 to 75 ft/min).

In International Standard units (note: 1 mm = 0.001 m and 1 L = 0.001 m³):

Calculate the minimum and maximum flowrates:

$$\text{Minimum flow} = \text{Flowrate} \times \text{Minimum number of units}$$

$$\text{Minimum flow} = 472 \text{ L/s} \times 2 \text{ units} = 944 \text{ L/s}$$

$$\text{Maximum flow} = \text{Flowrate} \times \text{Maximum number of units}$$

$$\text{Maximum flow} = 472 \text{ L/s} \times 3 \text{ units} = 1416 \text{ L/s}$$

Calculate the surface area at maximum flow:

$$\text{Surface area} = \text{Maximum flowrate /Maximum face velocity}$$

$$(1416 \text{ L/s} \times 0.001 \text{ m}^3/\text{L})/(380 \text{ mm/s} \times 0.001 \text{ m/mm}) = 3.73 \text{ m}^2$$

Calculate the surface area at minimum flow:

$$\text{Surface area} = \text{Minimum flowrate/Minimum face velocity}$$

In U.S. customary units:

Calculate the minimum and maximum flowrates:

$$\text{Minimum flow} = \text{Flowrate} \times \text{Minimum number of units}$$

$$\text{Minimum flow} = 1000 \text{ cfm} \times 2 \text{ units} = 2000 \text{ cfm}$$

$$\text{Maximum flow} = \text{Flowrate} \times \text{Maximum number of units}$$

$$\text{Maximum flow} = 1000 \text{ cfm} \times 3 \text{ units} = 3000 \text{ cfm}$$

Calculate the surface area at maximum flow:

Surface area = Maximum flowrate/Maximum face velocity

3000 cfm/75 ft/min = 40 sq ft

Calculate the surface area at minimum flow:

Surface area = Minimum flowrate/Minimum face velocity

Because the maximum flow is the limiting parameter, the minimum surface area should be 3.73 sq m (40 sq ft).

SOLUTION 13.11

Five pieces of information that should be recorded on the complaint form and explanations of how that information would/could be used include any of the following:

- Contact information for the person making the complaint. This can be used for follow-up and also would be useful for tracking the number of complaints from the individual over time.
- Location where odor was detected. This information is necessary for follow-up investigation and may be used to help locate the source of the odor.
- Time and date odor was detected and duration of odor. This information is critical for follow-up investigation and researching potential sources within your facility. Note that people often delay reporting, sometimes for several days.
- Weather conditions at the time the odor was detected, including wind direction, wind speed, and temperature. This information is critical in determining the source of the odor and could indicate that the facility is not the source.
- A description of the odor characteristics (smells like) and intensity. This can help narrow down the likely sources within the facility.
- The date and time the complaint was received. This is important because it can help explain the response taken, especially if it is substantially after the date and time the odor was detected.
- The name of the person filling out the form. This is important, especially if the facility has multiple shifts and the person following up is on a different shift schedule from the person taking the complaint.
- On-site inspection data, including odors detected, weather conditions, activity observed in the vicinity of the observed odor, and pictures if appropriate. This information can help in researching the odor source and defining the effect of the odor on local residents.

- Any abnormal operating conditions (especially if associated with units that could account for the described odor) at the facility at the time the odor was detected. This information may help decide if this is an ongoing issue or one that is of short duration and atypical of normal facility operation.

SOLUTION 13.12

g)   Options b), c), and e) should be considered

Given the physical observations, the first step would likely be to increase the rate of sludge pumping because it appears that the sludge is going septic. Additionally, the hydraulic detention time appears to be too high, especially given the influent dissolved oxygen and temperature. Thus, removing one or more clarifiers will reduce the potential for production of odorous compounds as a result of anaerobic activity. The addition of chlorine or iron compounds to the clarifier influent will help to delay the development of anaerobic activity and/or will reduce the amount of reduced sulfur compounds available for release.

Chapter 14
INTEGRATED PROCESS MANAGEMENT

Problems

PROBLEM 14.1

A biological treatment system is the only part of a water resource recovery facility (WRRF) that can be disrupted by operational changes from chemical and/or physical treatment systems.

a) True
b) False

PROBLEM 14.2

Downstream process performance cannot affect an upstream process.

a) True
b) False

PROBLEM 14.3

List two ways that operation of chemical and physical processes can affect a biological process.

PROBLEM 14.4

List two ways that operation of physical and biological processes can affect a chemical process.

PROBLEM 14.5

List two ways that operation of chemical and biological processes can affect a physical process.

PROBLEM 14.6

The standard operating procedures (SOPs) format should encourage new ideas and creative solutions. The "we have always done it that way" syndrome should be avoided.

a) True
b) False

PROBLEM 14.7

What are the four primary areas of concern in a process control management plan?

a) Assigning responsibilities for individual treatment processes, identifying technical support resources, establishing information references, and providing for redundancy
b) Establishing a chain of command for emergency problems only, assigning sampling and testing protocols to shift foremen, and sending only managers to process control seminars to save money
c) Lacking redundancy because of added costs, securing all maintenance and operations manuals so that operators do not have access to prevent loss of these materials, and hiring a consultant to maintain all process control records

PROBLEM 14.8

What are some effects that a poorly performing solids process can have on facility operations?

PROBLEM 14.9

List three main elements required for implementing an integrated process management strategy.

PROBLEM 14.10

Five-day biochemical oxygen demand test results are much too late for most daily process decisions.

a)   True
b)   False

PROBLEM 14.11

A nitrifying activated sludge facility has to have a basin taken out of service for cleaning. What are potential effects of taking a basin out of service for cleaning on the final effluent chlorination process? The aeration tank has two basins and is operated in parallel. The facility is operating close to capacity.

PROBLEM 14.12

The supernatant from an anaerobic digester is recycled to the primary clarifiers at a flowrate of 75 700 L/d (20 000 gpd). Because of a mechanical malfunction, the supernatant has a concentration of 4000 mg/L of total suspended solids (TSS) and 1000 mg/L biochemical oxygen demand (BOD). The primary clarifiers remove 30% of the influent BOD and 40% of the influent TSS. The aeration tank contains 5443 kg (12 000 lb) of mixed liquor suspended solids (MLSS) at 79% volatile. How many additional kilograms (pounds) of TSS will enter the solids processing system? If the mechanical problem is not corrected and the supernatant continues to have high concentrations of BOD and TSS, what would be the volatile solids content in the aeration tanks?

PROBLEM 14.13

For the previous problem, what would be the annual cost to supply air to the aeration tank if the additional BOD is to be completely removed from the secondary effluent? The oxygen consumption is 1.12 kg O_2/kg BOD removed and the cost to transfer air from the aeration blower is $0.082/kg O_2 ($0.037/lb O_2).

PROBLEM 14.14

What are some other potential effects from the supernatant on WRRF performance?

PROBLEM 14.15

Which of the following are needed for good data collection?

a) Pump running time, proper sampling equipment, and grab samples

b) Flow measurement and quantification, analytical analysis, maintenance reports, and process control data

c) Chain-of custody forms, 100-mL plastic bottles for all samples, and no special sampling devices

d) Non-flow-proportioned composite samples

PROBLEM 14.16

List the sources of data that would be important for generating an integrated process management plan for correcting a poorly performing conventional activated sludge facility (the secondary system is having settling problems in the secondary clarifiers).

PROBLEM 14.17

What four steps are essential to manage any process?

PROBLEM 14.18

The objective of process and instrumentation drawings is to account for all solids or flow that enter and leave a given process.

a) True

b) False

PROBLEM 14.19

Mass solids balance may be used on one process, such as thickening, or it may be used on a facility-wide basis.

a) True

b) False

PROBLEM 14.20

List some common types of graphs.

Solutions

SOLUTION 14.1

b)   False

Physical and chemical treatment systems can also be disrupted.

SOLUTION 14.2

b)   False

Most WRRFs contain recycle flow that potentially carries the effects of a poorly performing downstream system to upset an upstream process.

SOLUTION 14.3

The following are two ways that operation of chemical and physical processes can affect a biological process:

- A biological system can become hydraulically overloaded by too high of a return rate or facility drain sump recirculation rate.
- Facility recycle systems with high chemical residuals can destroy favored treatment bacteria.

SOLUTION 14.4

The following are two ways that operation of physical and biological processes can affect a chemical process:

- Poor settling in secondary clarifiers or poor filter operation in tertiary facilities can reduce disinfection effectiveness.
- Incomplete nitrification in biological processes can consume excessive amounts of chlorine as a result of nitrite in the secondary effluent.

SOLUTION 14.5

The following are two ways that operation of chemical and biological processes can affect a physical process:

- Poorly performing biological systems can inhibit the settling in secondary clarifiers, which will lead to a declining secondary effluent quality.

- Excessive prechlorination can diminish physical straining capabilities in single-stage media filters. Over-coagulation can produce high head loss and reduce the hydraulic loading of the tertiary filters.

SOLUTION 14.6

a) True

SOLUTION 14.7

a) Assigning responsibilities for individual treatment processes, identifying technical support resources, establishing information references, and providing for redundancy

SOLUTION 14.8

The following are some effects that a poorly performing solids process can have on facility operations:

- Improperly operated solids processes can recycle supernatants high in BOD and/or ammonia-nitrogen to the headworks and primary systems, increasing the organic load to the secondary system. Higher oxygen demand and ammonia level increase energy costs associated with increased air requirements.
- Thinner sludge sent to digestion will result in an inefficient use of heat (anaerobic) or air (aerobic), resulting in an increased value of energy cost/volatile solids reduced.
- Poor dewatering operation greatly increases sludge transportation/application costs by having to transport a greater volume of sludge as a result of a lower mass-to-volume ratio (concentration) of solids.
- Recycled supernatants high in colloidal and dissolved solids have the potential of degrading effluent quality and exceeding permit limits.

SOLUTION 14.9

The following are three main elements required for implementing an integrated process management strategy:

- Development and implementation of a process control plan,
- Review and evaluation of process performance, and
- Assessment of changes in process control strategies.

SOLUTION 14.10

a)   True

SOLUTION 14.11

The following are potential effects of taking a basin out of service for cleaning on the final effluent chlorination process:

- Taking a basin out of service decreases the mean cell residence time by 50%. The reduction in residence time may limit the BOD conversion in the aeration tank, leading to decreased effluent quality.

- The volume of activated sludge in the emptied basin will likely recycle back to the beginning of the facility, placing an increased oxygen demand on the remaining basin in service, further reducing BOD conversion within the basin.

- An incomplete conversion of nitrification will produce nitrite as an intermediary step between the conversion of ammonia to nitrate. Nitrite in the secondary effluent will exert a chlorine demand of 5 mg/L for every 1 mg/L concentration of nitrite.

- Ammonia present in unnitrified secondary effluent will react with chlorine, creating a chloramine residual unexpected on the breakpoint curve. Sufficient chlorine must be added until the residual can move past the breakpoint and into a "free chlorine" state.

SOLUTION 14.12

121 kg/d (266 lb/d) and 76.4% volatile

In International Standard units:

Determine the amount of solids added to the primary system:

$$(75\ 700\ \text{L/d}) \times (4000\ \text{mg/L}) \times (1\ \text{kg}/1000\ \text{g}) \times (0.001\ \text{g}/1\ \text{mg}) = 303\ \text{kg TSS/d}$$

Determine the amount of solids added to the secondary system:

$$(1 - 0.4) \times (303\ \text{kg/d}) = 182\ \text{kg TSS/d}$$

The solids-handling system must now unnecessarily transport an additional 121 kg/d of solids.

The solids from the digester are considered inert (digested), and the new percent volatile in the secondary system is determined as follows:

$$(5443 \text{ kg}) \times (0.79)/(5443 \text{ kg} + 182 \text{ kg}) = 76.5\% \text{ volatile}$$

In U.S. customary units:

Determine the amount of solids added to the primary system:

$$20\,000 \text{ gpd} \times (8.34 \text{ lb/mil. gal/mg/L}) \times (4000 \text{ mg/L}) = 667 \text{ lb TSS/d}$$

Determine the amount of solids added to the secondary system:

$$(1 - 0.4) \times (667 \text{ lb/d}) = 400 \text{ lb TSS/d}$$

The solids-handling system must now unnecessarily transport an additional 266 lb/d of solids.

The solids from the digester are considered inert (digested), the new percent volatile in the secondary system is determined as follows:

$$(12\,000 \text{ lb}) \times (0.79)/(12\,00 \text{ lb} + 400 \text{ lb}) = 76.5\% \text{ volatile}$$

SOLUTION 14.13

$1800 per year

In International Standard units:

Additional amount of BOD added to the secondary system:

$$(75\,700 \text{ L/d}) \times (1 \text{ kg}/1000 \text{ g}) \times (0.001 \text{ g}/1 \text{ mg}) \times (1000 \text{ mg/L}) \times (1 - 0.3) = 53 \text{ kg BOD/d}$$

Additional annual cost for supernatant BOD loading:

$$(1.12 \text{ kg/kg}) \times (\$0.082/\text{kg}) \times (53 \text{ kg/d}) \times (365 \text{ d/yr}) = \$1776/\text{yr (round to } \$1800/\text{yr)}$$

In U.S. customary units:

Additional amount of BOD added to the secondary system:

$$20\,000 \text{ gpd} \times (8.34 \text{ lb/mil. gal/mg/L}) \times (1000 \text{ mg/L}) \times (1 - 0.3) = 117 \text{ lb BOD/d}$$

Additional annual cost for supernatant BOD loading:

$$(1.12 \text{ lb/lb}) \times (\$0.037/\text{lb}) \times (117 \text{ lb/d}) \times (365 \text{ d/yr}) = \$1770/\text{yr (round to } \$1800/\text{yr)}$$

SOLUTION 14.14

The following are some other potential effects from the supernatant on WRRF performance:

- Additional inert solids from the supernatant appear as MLSS if a volatile solids test is not performed. Using only MLSS can lead to overwasting while trying to maintain a target mean cell residence time, reducing the active mixed liquor volatile suspended solids available for BOD reduction, and raising the food-to-microorganism ratio, causing poor settling and reduction of effluent quality.
- Additional oxygen demand can take a nitrifying facility out of nitrification, leading to changes in disinfection and effluent quality.
- Digester supernatant can contain high colloidal and dissolved solids that can move through the WRRF and decrease effluent quality.
- Digester capacity is decreased as a result of recirculation of uncaptured solids, lowering the potential contact time for volatile solids reduction.

SOLUTION 14.15

b)   Flow measurement and quantification, analytical analysis, maintenance reports, and process control data

SOLUTION 14.16

The following are sources of data that would be important for generating an integrated process management plan for correcting a poorly performing conventional activated sludge facility (the secondary system is having settling problems in the secondary clarifiers):

- Influent, return activated sludge, and waste activated sludge flow trends;
- Mixed liquor volatile suspended solids, influent and secondary effluent BOD trends, pH, temperature trend, alkalinity, and ratios of indicator organisms in the MLSS;
- Pump maintenance records and run hours, chemical feed maintenance records for aeration buffer agent, diffuser and blower maintenance records, and calibration records for sensors; and
- Sludge volume index and mean cell residence time trends and food-to-microorganism ratio.

SOLUTION 14.17

The following four steps are essential to manage any process

- Gather information,
- Evaluate the data,
- Develop and implement a proper response,
- Reevaluate.

SOLUTION 14.18

b) False

SOLUTION 14.19

a) True

SOLUTION 14.20

The following are some common types of graphs:

- Line,
- Bar,
- Scatter, and
- Pie chart.

Chapter 15
OUTSOURCED OPERATIONS SERVICES AND PUBLIC/ PRIVATE PARTNERSHIPS

Problems

PROBLEM 15.1

Which is not a common type of outsourced operation service?

   a)   Design, build, operate
   b)   Public/labor union partnership
   c)   Full privatization
   d)   Merchant facility
   e)   None of the above

PROBLEM 15.2

Outsourced operation services are the result of the need to combat the local effects of global warming.

   a)   True
   b)   False

PROBLEM 15.3

Operation services contractors are limited, by Internal Revenue Service Procedure 97-13, to perform either "inside-the-fence" or "outside-the-fence" contracts, but not both.

   a)   True
   b)   False

PROBLEM 15.4

The size of the facility and/or system has no effect on the public/private partnership contractor's ability to improve the overall facility cost savings.

   a)   True
   b)   False

PROBLEM 15.5

The facility's location (remote vs urban) can influence the ability of the service provider to achieve the necessary economies of scale in management and operational areas to produce material cost savings.

   a)   True
   b)   False

PROBLEM 15.6

The corporate experience of the operation services vendor is the most important criterion for the owner to consider during the proposal stage of contract development and execution.

   a)   True
   b)   False

PROBLEM 15.7

Steps to "short list'" the operation services vendor during the proposal stage of contract development and execution do not include which of the following?

   a)   Project-related experience
   b)   Client reference
   c)   Quantity of the submitted proposal drawings
   d)   Project costs
   e)   All of the above

PROBLEM 15.8

While no single evaluation method is universally accepted, a numerical matrix that assigns points to the various evaluation parameters is typical.

a) True

b) False

PROBLEM 15.9

Facility employee issues are a factor for unionized employees and may affect the success or failure of an outsourcing contract.

a) True

b) False

PROBLEM 15.10

An "unlimited liability" proposal requirement from the operation services vendor is significantly better than a "bonded guarantee".

a) True

b) False

PROBLEM 15.11

The terms for a contract buyout must recognize the value of

a) expenditures associated with project startup.

b) ongoing operations.

c) lost opportunity for the contractor.

d) All of the above

PROBLEM 15.12

In the water and wastewater utility business, there are currently

a) greater than 10% of the facilities operating under a public/private partnership agreement.

b) 1 to 3% of the facilities operating under a public/private partnership agreement.

c) 3 to 5% of the facilities operating under a public/private partnership agreement.

d) 5 to 7% of the facilities operating under a public/private partnership agreement.

PROBLEM 15.13

The operation services contractor industry may be divided into the following categories:

Solutions

SOLUTION 15.1

b) Public/labor union partnership

SOLUTION 15.2

b) False

The effects of the requirement of the Clean Water Act, the Safe Drinking Water Act, and other environmental regulatory requirements may be the driving forces for public/private partnership services.

SOLUTION 15.3

b) False

Revenue Procedure 97-13 actually expanded the public/private partnership contractual capabilities and contractual time duration.

SOLUTION 15.4

b) False

The opportunity for cost savings by the operation services contractors depends on several conditions, including the size of the facility and/or system.

SOLUTION 15.5

a) True

SOLUTION 15.6

a) True

SOLUTION 15.7

c) Quantity of the submitted proposal drawings

SOLUTIONS 15.8

a) True

SOLUTION 15.9

b) False

Resolution of employee issues, without regard to union or nonunion status, is a critical factor regarding the success or failure of an outsourcing contract.

SOLUTION 15.10

b) False

The "unlimited liability" proposal requirement by the operation services vendor is unreasonable and not beneficial to the owner. A "bonded guarantee" is significantly better for the owner because it has real and intrinsic value.

SOLUTION 15.11

d) All of the above

SOLUTION 15.12

c) 3 to 5% of the facilities operating under a public/private partnership agreement (as of 2007).

SOLUTION 15.13

The operation services contractor industry may be divided into the following categories:

- International/national-level firms,
- Regional/national-level firms, and
- Regional/local-level firms.

Chapter 16
TRAINING

Problems

PROBLEM 16.1

What is training?

a) Training is required to satisfy the directives provided to management by legal staff.

b) Training is a system that provides information that people need to do the work of the organization.

c) Training is the presentation of a set curricula using prespecified methods to groups of employees.

PROBLEM 16.2

An organization that has a sizeable training budget

a) will never have any problems training and retaining qualified staff.

b) can have managerial stresses, poor job performance, and frequent breakdowns in its physical facility if it does not provide support to its training teams.

c) should be sure to fully expend its training budget each year to maintain funding support.

PROBLEM 16.3

A trainer should be selected using the following criteria:

a) The trainer should have more years of experience than any of the staff assigned to receive the training

b) The trainer should be someone who is not busy, to avoid interrupting more important work activities.

c) The trainer should be proficient at the specific job, communicate effectively, and know how to effectively organize a presentation and adapt it to the trainees' abilities

PROBLEM 16.4

Training will succeed if

a) the training covers multiple topics in one session to reduce the overall time lost from regularly scheduled work activities.

b) the training duplicates previous sessions to ensure that trainees retain the information presented.

c) the training covers new material, on a single topic, that is relevant to the trainees' needs for skill and knowledge.

PROBLEM 16.5

A large agency with 600 employees estimates that it will require 8000 work hours during an initial 1-year period to prepare a consolidated set of written procedures for facility operations. Employees at the agency work an average of 1840 hours per year. Once completed, the procedures need to be updated again in 5 years. What percentage of the agency's 5-year total labor effort is required for preparation of the consolidated set of written procedures for facility operations?

PROBLEM 16.6

Name at least five topics that could be incorporated to a comprehensive safety-training program at a wastewater agency.

PROBLEM 16.7

Renewal of operator certification depends on the class of the license, and varies from state to state. As part of a specific state-operator certification-training program, an operator is required to document at least 20 professional development hours during each 2-year period to renew his/her operator's license. The operator attends a 3-day conference that provides a total of 12 technical program sessions and two workshops. The operator is able to attend one workshop providing 4.5 professional development hours and six technical program sessions, each technical program session providing 2.5 professional development hours. How many professional development hours are earned by the operator attending the conference? What percentage of the overall professional development hours has the operator earned by attending the conference?

PROBLEM 16.8

What is a policy?

a) A policy is an important document that is used by management to form the basis for disciplinary action.

b) A policy is a brief and vague administrative statement providing general information used to control the activities of staff.

c) A policy is an administrative document governing a single element of a facility, department, or work group outlining management and workplace practices.

PROBLEM 16.9

Written procedures should incorporate which of the following?

a) Conditional language such as "could", "should", "may", or "might" that is used to place responsibility on the employee for their actions

b) Procedures written in sequential order, in an extended outline form, including safety-related considerations, and updated on an ongoing basis

c) Procedures that encourage employee creativity while performing important work tasks to minimize boredom

PROBLEM 16.10

Operating manuals are a compilation of information about equipment and processes containing all policies and procedures necessary for both normal and irregular operation. Operating manuals should

a) be brief so that valuable field time is not consumed learning about a task.

b) address all topics needed to provide a complete operational guide for the equipment or the process.

c) not cover safety-related matters, which should be the subject of another separate, stand-alone document.

PROBLEM 16.11

Which of the following statements about training is correct?

a) Staff should be trained in response to clear and immediate need and as a substitute for supervisory remedies.

b) Training must be conducted with complete independence to provide objectivity.

c) Training must be practical and use methods that are appropriate for the subject matter.

PROBLEM 16.12

Formal verification of training should be documented because

a) legal staff will need this if something goes wrong after the training is performed to avoid serious financial and criminal liabilities.

b) it provides supervisory personnel with positive documentation that staff and crew are now experts.

c) it provides a written record of attendance to avoid repeating classes and allows supervisors to identify employees who have not received the required training.

PROBLEM 16.13

Adults learn for different reasons than children, and learn in different ways. While a child may learn for the joy of it, most adults do not.

a) True

b) False

PROBLEM 16.14

Demonstrating new skills or knowledge gives an individual immediate confidence and a strong impetus to use them.

a) True

b) False

PROBLEM 16.15

When analyzing a potential training package, trainers should first assess the level of mastery required to do the portion of the job covered by the training.

a) True

b) False

PROBLEM 16.16

List some of the most common and serious pitfalls that can harm, or doom, a training program.

Solutions

SOLUTION 16.1

b)   Training is a system that provides information that people need to do the work of the organization.

SOLUTION 16.2

b)   can have managerial stresses, poor job performance, and frequent breakdowns in its physical facility if it does not provide support to its training teams.

SOLUTION 16.3

c)   The trainer should be proficient at the specific job, communicate effectively, and know how to effectively organize a presentation and adapt it to the trainees' abilities.

SOLUTION 16.4

c)   the training covers new material, on a single topic, that is relevant to the trainees' needs for skill and knowledge.

SOLUTION 16.5

14%

SOLUTION 16.6

Five topics that could be incorporated to a comprehensive safety-training program at a wastewater agency include any of the following:

- Safety and health policies;
- Basic safety and health for wastewater professionals;
- First aid and cardiopulmonary resuscitation;
- Wastewater hazards;
- Industrial hygiene;
- Respiratory protective equipment;
- Hand and power tool safety;
- Fire prevention and control;

- Accident and illness reporting;
- Accident investigation;
- Hazardous energy control (lockout/tagout) procedures;
- Confined-space entry;
- Safe work permitting procedures;
- Emergency response;
- Safety for hydrogen sulfide and other gases;
- Chemical handling;
- Hazard communication and labeling (material safety data sheets);
- Hearing, eye, and face protection;
- Right-to-know laws for employees and the community;
- Housekeeping and safe equipment storage; and
- Electrical hazards, including hot work and basic awareness.

SOLUTION 16.7

19.5 hours, which is 97.5%

$$4.5 \text{ hours} = (6 \text{ hours} \times 2.5 \text{ hours}) = 19.5 \text{ hours}$$

$$(19.5 \text{ hours}/20 \text{ hours}) \times 100\% = 0.975, \text{ or } 97.5\%$$

SOLUTION 16.8

c) A policy is an administrative document governing a single element of a facility, department, or work group outlining management and workplace practices.

SOLUTION 16.9

b) Procedures written in sequential order, in an extended outline form, including safety-related considerations, and updated on an ongoing basis

SOLUTION 16.10

b) address all topics needed to provide a complete operational guide for the equipment or the process.

SOLUTION 16.11

c)   Training must be practical and use methods that are appropriate for the subject matter.

SOLUTION 16.12

c)   it provides a written record of attendance to avoid repeating classes and allows supervisors to identify employees who have not received the required training.

SOLUTION 16.13

a)   True

SOLUTION 16.14

a)   True

SOLUTION 16.15

a)   True

SOLUTION 16.16

Some of the most common and serious pitfalls that can harm, or doom, a training program include the following:

- Duplication,
- Repetition,
- Scattershot,
- Failing to lead,
- Substitution for managerial remedies,
- Too early or too late, and
- Unnecessary.

Chapter 17
CHARACTERIZATION AND SAMPLING OF WASTEWATER

Problems

PROBLEM 17.1

Volatile solids provide a good approximation of how much organic matter is present in the wastewater.

- **a)** True
- **b)** False

PROBLEM 17.2

Samples that are taken at equal time intervals with the sample volume varied based on the flowrate at the moment of sampling are

- **a)** Grab samples.
- **b)** Flow-proportioned composite samples.
- **c)** Time composite samples.

PROBLEM 17.3

pH and alkalinity are the same thing.

- **a)** True
- **b)** False

PROBLEM 17.4

Before entering any confined space, test the atmosphere to ensure a safe atmosphere that has sufficient oxygen levels, no or low levels of hydrogen sulfide, and no evidence of which of the following?

a) Hydrogen dioxide
b) Carbon dioxide
c) Explosivity

PROBLEM 17.5

Phosphorus-accumulating organisms are cultivated for which of the following?

a) Anaerobic digestion
b) Conventional activated sludge process
c) Biological nutrient removal

PROBLEM 17.6

Under acidic conditions, polyphosphates (which are hydrolysable compounds) can be converted to which of the following?

a) Organic phosphorus
b) Orthophosphate
c) Monophosphate

PROBLEM 17.7

Low-density fats, oils, and grease (FOG)

a) are not removed in the secondary system and will pass out of the facility in the effluent.
b) may create poor settleability in the secondary system.
c) enhance activated sludge settleability by promoting flocculation.

PROBLEM 17.8

The activated sludge 30-minute settleability test indicates 680-mL settling in a 2-L graduated cylinder. If the aeration tank mixed liquor suspended solids concentration is 2275 mg/L, calculate the sludge volume index (SVI).

PROBLEM 17.9

A water reclamation facility uses chemical oxygen demand (COD) to predict effluent biochemical oxygen demand (BOD). The laboratory has recently updated the correlation between the two. Using the chart below, predict the 5-day BOD (BOD_5) value for a COD of 83 mg/L.

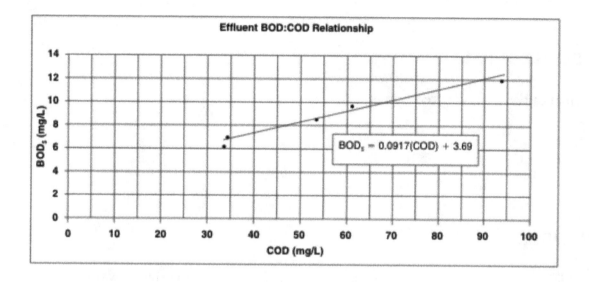

PROBLEM 17.10

A water reclamation facility operates the activated sludge system to biologically remove nitrogen. Alkalinity is produced in the anoxic zone, which precedes the aerobic zone. Oxidation–reduction potential (ORP) sensors are located in both zones. Currently, the sensors indicate an ORP of 1150 mV in the aerobic zone and 2150 mV in the anoxic zone. Based on the information depicted in the table, select the answer that best describes the performance.

Anaerobic Fermentation	−200 mV to −50 mV
Anoxic Denitrification	−50 mV to +50 mV
Aerobic Carbonaceous BOD	+50 mV to +225 mV
Nitrification	+100 mV to +325 mV

a)   The system is working well for both nitrification and denitrification.

b)   The ORP is too high in the denitrification tanks, which indicates incomplete or no denitrification (which may result in loss of alkalinity).

c)   A filamentous problem may be present because of low ORP in the denitrification tanks.

d)   Nitrification is not taking place because the ORP is too low.

PROBLEM 17.11

The nitrification process will increase pH and regain alkalinity loss during the denitrification process.

a) True
b) False

PROBLEM 17.12

Which of the following relation is correct?

a) $^{\circ}C = 9/5\ (^{\circ}F - 32)$
b) $^{\circ}C = 5/9\ (^{\circ}F - 32)$
c) $^{\circ}F = 9/5\ (^{\circ}C - 32)$

PROBLEM 17.13

Colder water is capable of holding more dissolved oxygen then warmer water.

a) True
b) False

PROBLEM 17.14

The FOG samples must be stored in a polyethylene bottle.

a) True
b) False

Solutions

SOLUTION 17.1

a) True

SOLUTION 17.2

b) Flow-proportioned composite samples.

SOLUTION 17.3

b) False

The term, *pH,* is traditionally used as a convenient representation of the concentration of hydrogen ions in a solution. Alkalinity is a measure of the ability of the wastewater to neutralize acid.

SOLUTION 17.4

c) Explosivity

SOLUTION 17.5

c) Biological nutrient removal

In a biological nutrient removal system, conditions are created to favor phosphorus-accumulating organisms. These organisms uptake excess phosphorus under aerobic conditions and release it under anaerobic conditions.

SOLUTION 17.6

b) Orthophosphate

Orthophosphate is the form most available to biota.

SOLUTION 17.7

b) may create poor settleability in the secondary system.

If excessive levels of FOG enter a secondary system, the low-density FOG constituents merge with the biomass. This merge can cause poor settleability of the biological solids, with a resultant excessive solids loss to the effluent.

SOLUTION 17.8

149 mL/g

$$\text{SVI, mL} = \frac{(680\ \text{mL})(1000\ \text{mg/g})}{(2\ \text{L})(2275\ \text{mg/L})} = 149\ \text{mL/g}$$

Note that good quality activated sludge has an SVI between 50 and 100 mL/g. As the SVI increases to 200 mL/g or more, the sludge is characterized as a "bulking sludge".

SOLUTION 17.9

11 mg/L BOD_5

The BOD:COD relation equation is as follows:

$$BOD_5 = 0.0917(COD) + 3.69$$

$$= 0.0917(83 \text{ mg/L}) + 3.69 = 11 \text{ mg/L } BOD_5$$

SOLUTION 17.10

b) The ORP is high in the denitrification tanks, which indicates incomplete or no denitrification (which may result in loss of alkalinity).

The denitrification tank ORP indicates aerobic conditions.

SOLUTION 17.11

b) False

The nitrification process will consume alkalinity and may lower pH.

SOLUTION 17.12

b) $°C = 5/9 \,(°F - 32)$

SOLUTION 17.13

a) True

Oxygen solubility increases with decreasing water temperature.

SOLUTION 17.14

b) False

The FOG samples must be stored in a glass bottle.

Chapter 18
PRELIMINARY TREATMENT

Problems

PROBLEM 18.1

Increased grit removal and excessive organic material in the grit chambers may adversely affect facility operation because

    a) there is no adverse effect. Anything that settles in the grit chamber should be removed.

    b) fewer activated sludge trains will need to be operated because less biochemical oxygen demand (BOD) is entering the system.

    c) landfills will refuse to accept grit because of odor or water content.

PROBLEM 18.2

Bar racks typically have a bar spacing between

    a) 50 and 150 mm (2 and 6 in.).

    b) 150 and 250 mm (6 and 9 in.).

    c) 12 and 50 mm (0.5 and 2 in.).

PROBLEM 18.3

During morning rounds, the operator notes that the aerated grit system output is less than normal for this time of day. One cause of decrease output could be that the

    a) air-to-grit chamber is insufficient.

    b) cyclone inlet pressure is too low.

    c) bucket elevator is moving too fast and the speed needs to be reduced.

PROBLEM 18.4

What is the purpose of spray nozzles on a grit washer?

a) To concentrate the grit for more effective removal
b) To remove stray putrescibles
c) To prevent grit from getting caught in screw auger bearings

PROBLEM 18.5

The stretch in a chain-driven bar screen chain is caused by which of the following?

a) Buildup on the drive sprocket has lengthened the chain
b) Hydrogen sulfide has corroded the chain
c) The diameter of the chain link barrel has decreased because of wear

PROBLEM 18.6

The bar screens are being replaced with rotary drum screens in the headworks. What change in activities should be adjusted with the new installation?

a) Schedule screenings transfer to the landfill more frequently.
b) Increase the preventive maintenance schedule.
c) Increase the chemical budget for caustic to headworks scrubbers.

PROBLEM 18.7

To be accepted at landfills, materials (including grit and screenings) typically must pass the paint filter test. What are the length of the test period and the volume of liquid that can pass through mesh no. 60 for this test?

a) 30 minutes and no liquid
b) 10 minutes and 5 mL of liquid
c) 5 minutes and no liquid

PROBLEM 18.8

A water reclamation facility operates the activated sludge system in the nitrification mode. Which of the following situations is most likely to affect operation of the secondary system?

a) Belt filter press may create high hydraulic loading as a result of filtrate and belt cleaning activities.

b) Dewatering return flows may cause shock loads.

c) Thickener flows may contain elevated levels of phosphorus resulting from a secondary release.

PROBLEM 18.9

Determine the screenings accumulation rate if a 6-m³ (8-cu-yd) container is collected in a 24-hour period with a recorded flow of 877 L/s (20 mgd).

PROBLEM 18.10

If short-circuiting occurs in an aerated grit chamber, an operator should consider which of the following:

a) Installing submerged transverse or longitudinal baffles

b) Increasing the airflow rate

c) Reducing the airflow rate

Solutions

SOLUTION 18.1

c) landfills will refuse to accept grit because of odor or water content.

SOLUTION 18.2

a) 50 and 150 mm (2 and 6 in.)

SOLUTION 18.3

b)   cyclone inlet pressure is too low.

Increase inlet pressure by adjusting the grit pump discharge.

SOLUTION 18.4

b)   To remove stray putrescibles

SOLUTION 18.5

c)   The diameter of the chain link barrel has decreased because of wear

SOLUTION 18.6

a)   Schedule screenings transfer to the landfill more frequently (before excess odors are produced).

SOLUTION 18.7

c)   5 minutes and no liquid

SOLUTION 18.8

b)   Dewatering return flows may cause shock loads.

Additional loads from digester decant or dewatering activities may overload the secondary system.

SOLUTION 18.9

0.08 m³/ML (0.4 cu yd/mil. gal)

In International Standard units:

$$\text{Screenings accumulation rate} = \frac{(\text{Screenings Collected, m}^3/\text{d}) \times (1\,000\,000\ \text{L/ML})}{(\text{Flow, L/s}) \times (60\ \text{s/min}) \times (1440\ \text{min/d})}$$

$$\frac{(6\ \text{m}^3/\text{d}) \times (1\,000\,000\ \text{L/ML})}{(900\ \text{L/s}) \times (60\ \text{s/min}) \times (1440\ \text{min/d})} = 0.08\ \text{m}^3/\text{ML}$$

In U.S. customary units:

$$\text{Screenings accumulation rate} = \frac{\text{Screenings collected, cu yd/d}}{\text{Flow, mgd}}$$

$$\frac{8 \text{ cu yd/d}}{20 \text{ mgd}} = 0.4 \text{ cu yd/mil. gal}$$

SOLUTION 18.10

a)   Installing submerged transverse or longitudinal baffles

Submerged transverse baffles or longitudinal baffles should be installed adjacent to the diffusers or along the wall opposite the diffusers.

Chapter 19
PRIMARY TREATMENT

Problems

PROBLEM 19.1

Suspended solids can be classified as _____ or _____.

PROBLEM 19.2

Calculate the surface overflow rate (SOR) of a clarifier, given the following: Q = 76 360 m³/d (20 mgd); number of units = 3; diameter = 40 m (130 ft); sidewall depth = 4 m (13 ft); and primary effluent biochemical oxygen demand (BOD) = 102 mg/L.

a) 60 m/d (1520 gpd/sq ft)
b) 20 m/d (503 gpd/sq ft)
c) 15 m/d (760 gpd/sq ft)
d) 8 m/d (58 gpd/sq ft)

PROBLEM 19.3

Name three variables that affect the removal of floatables in a primary sedimentation tank:

a) Tank configuration, overflow rate, and solids loading rates
b) Tank configuration, primary effluent characteristics, and removal efficiency
c) Tank configuration, wastewater characteristics, and removal mechanism design and configuration
d) Tank configuration, sidestream flow and loads, and sludge scraper design

PROBLEM 19.4

Fine screens are considered primary treatment.

a) True
b) False

PROBLEM 19.5

The relationship between hydraulic retention time (HRT) and solids removal is described by which of the following:

a) The higher the hydraulic retention time, the lower the solids removal
b) The lower the solids retention time, the lower the solids removal
c) The lower the hydraulic retention time, the higher the solids removal
d) The higher the solids retention time, the higher the solids removal

PROBLEM 19.6

The facility has three primary clarifiers. The HRT at average flow is 3.5 hours. Which of the following should be done?

a) Decrease the flow coming into the facility by intermittently shutting down pumping stations.
b) Decrease the rate at which sidestreams enter the primary tanks.
c) Decrease the number of tanks in service.
d) Decrease the primary sludge pumping rate.

PROBLEM 19.7

Chemicals are never used in the primary sedimentation process.

a) True
b) False

PROBLEM 19.8

Calculate the percent removal of BOD and total suspended solids (TSS), respectively, given the following: influent BOD = 189 mg/L; primary effluent BOD = 110 mg/L; influent TSS = 200 mg/L; and primary effluent TSS = 60 mg/L.

a) 34 and 80%, respectively
b) 34 and 60%, respectively
c) 42 and 80%, respectively
d) 42 and 70%, respectively

PROBLEM 19.9

Calculate the HRT, given the following: Q = 107 100 m³/d (28 mgd), number of settling tanks = 2, R = 20 m (66 ft), and depth = 4 m (13 ft).

a) 2.25 hours
b) 4.5 hours
c) 0.09 hours
d) 2.25 days

PROBLEM 19.10

Average surface overflow rates (SORs) for primary sedimentation tanks are which of the following?

a) 122 m/d (3000 gpd/sq ft)
b) 32 to 49 m/d (800 to 1200 gpd/sq ft)
c) 32 to 49 gpd/sq ft (800 to 1200 m/d)
d) 122 kg/m²·d (3000 lb/d/sq ft)

PROBLEM 19.11

Primary sedimentation is a chemical process.

a) True
b) False

PROBLEM 19.12

A hypothetical water resource recovery facility has three primary clarifiers. Each one has a diameter of 15 m (50 ft), with a sidewater depth of 3.7 m (12 ft). Each clarifier receives a flow of 60 567 m³/d (16 mgd). The influent BOD is 205 mg/L and the influent TSS is 200 mg/L. The primary effluent BOD is 135 mg/L and the primary effluent TSS is 78 mg/L. What is the volume of each tank?

a) 521 m³ (159 198 gal)
b) 521 L (1.19 mil. gal)
c) 653 m³ (176 154 gal)
d) 654 L (2.18 mil. gal)

PROBLEM 19.13

A primary sludge pump ran continuously for 24 hours and delivered 1090 m³ (288 000 gal). The capacity of the pump is which of the following?

a) 45 L/d (100 gpm)
b) 760 000 L/d (200 gpm)
c) 30 m³/d (1000 gpm)
d) 45 m³/h (12 000 gph)

PROBLEM 19.14

Draw a simple mass balance diagram around a primary sedimentation tank. Only consider influent and primary effluent TSS and primary sludge.

Solutions

SOLUTION 19.1

Suspended solids can be classified as <u>flocculent</u> or <u>granular</u>.

SOLUTION 19.2

b) 20 m/d (503 gpd/sq ft)

In International Standard units:

$$\text{Area} = 3.14\,(20 \text{ m})^2 = 1256 \text{ m}^2$$

$$\text{SOR} = (76\,360 \text{ m}^3/\text{d})/(3 \times 1256 \text{ m}^2) = 20 \text{ m/d}$$

In U.S. customary units:

$$\text{Area} = 3.14(65 \text{ ft})^2 = 13\,266 \text{ sq ft}$$

$$\text{SOR} = (20\,000\,000 \text{ gpd})/(3 \times 13\,266 \text{ sq ft}) = 503 \text{ gpd/sq ft}$$

SOLUTION 19.3

c)   Tank configuration, wastewater characteristics, and removal mechanism design and configuration

SOLUTION 19.4

a)   True

SOLUTION 19.5

d)   The higher the solids retention time, the higher the solids removal

SOLUTION 19.6

c)   Decrease the number of tanks in service.

SOLUTION 19.7

b)   False

SOLUTION 19.8

d)   42 and 70%, respectively

SOLUTION 19.9

a)   2.25 hours

In International Standard units:

$$HRT = Volume/flow \ (V/Q)$$

$$Volume = 3.14 \ (20 \ m)^2 \times 4 \ m = 5024 \ m^3 \ per \ clarifier$$

$$HRT = \frac{5024 \ m^3 \times 2}{107 \ 100 \ m^3/d} = 24 \ h/d = 2.25 \ h$$

In U.S. customary units:

$$HRT = Volume/Flow \ (V/Q)$$

$$Volume = 3.14 \ (66 \ ft)^2 \times 13 \ ft = 177 \ 812 \ cu \ ft \times 7.48 \ gal/cu \ ft =$$
$$1 \ 330 \ 033 \ gal \ per \ clarifier$$

$$HRT = \frac{1 \ 330 \ 033 \ gal \times 2}{28 \ 000 \ 000 \ gpd} \times 24 \ hr/d = 2.28 \ hr \ (round \ to \ 2.25 \ hr)$$

SOLUTION 19.10

b)   32 to 49 m/d (800 to 1200 gpd/sq ft)

SOLUTION 19.11

b)   False

SOLUTION 19.12

c)   653 m³ (176 154) gal

In International Standard units:

$$\text{Volume} = 3.14 \, (7.5 \text{ m})^2 \times 3.7 \text{ m} = 653 \text{ m}^3$$

In U.S. customary units:

$$\text{Volume} = 3.14 \, (25 \text{ ft})^2 \times 12 \text{ ft} = 23\,550 \text{ cu ft} \times 7.48 \text{ gal/cu ft} = 176\,154 \text{ gal}$$

SOLUTION 19.13

d)   45 m³/h (12 000 gph)

In International Standard units:

$$\text{Capacity} = 1090 \text{ m}^3/24 \text{ h} = 45 \text{ m}^3/\text{h}$$

In U.S. customary units:

$$\text{Capacity} = 288\,000 \text{ gal}/24 \text{ hr} = 12\,000 \text{ gph}$$

SOLUTION 19.14

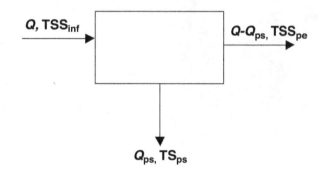

Where

Q = influent flow,

Q_{ps} = primary sludge flow,

TSS_{inf} = influent TSS concentration,

TSS_{pe} = primary effluent TSS concentration, and

TS_{ps} = primary sludge solids concentration.

Chapter 20
ACTIVATED SLUDGE

Problems

PROBLEM 20.1

List the basic system components of a conventional activated sludge process.

PROBLEM 20.2

The activated sludge process consists of flocculated microorganisms maintained in suspension by aeration and mixing.

- **a)** True
- **b)** False

PROBLEM 20.3

Microscopic techniques to identify filamentous organisms in activated sludge have been useful in which of the following?

- **a)** Diagnosing and controlling bulking problems
- **b)** Designing the activated sludge process
- **c)** Establishing the metabolic characteristics of filaments
- **d)** Have not been useful to facility operators

PROBLEM 20.4

Chemotrophs are which of the following?

   a) Organisms that use sunlight as their energy source
   b) Organisms that use oxygen as their energy source
   c) Organisms that use chemical compounds as their energy source
   d) Organisms that use enzymes as their energy source

PROBLEM 20.5

Autotrophs are organisms that get their energy from which of the following?

   a) Organic carbon
   b) Ammonia and nitrite
   c) Organic nitrogen
   d) Orthophosphate

PROBLEM 20.6

Heterotrophs use _____ as their carbon source.

PROBLEM 20.7

In aerobic reactions, the final electron acceptor is which of the following?

   a) Nitrate
   b) Ammonia
   c) Oxygen
   d) Nitrite

PROBLEM 20.8

The reaction that produces the most biomass is which of the following?

   a) Anaerobic
   b) Anoxic
   c) Aerobic
   d) Facultative

PROBLEM 20.9

What is the name of the metabolic function that reduces total biomass yield as the solids retention time (SRT) increases?

PROBLEM 20.10

Name four environmental conditions that are important in the activated sludge process.

PROBLEM 20.11

What does a selector do?

PROBLEM 20.12

An anoxic zone is considered a selector.

- **a)** True
- **b)** False

PROBLEM 20.13

Biological conditions associated with poor separation problems in secondary clarifiers include which of the following?

- **a)** Dispersed growth, pin floc, and bulking
- **b)** Short-circuiting and density currents
- **c)** Nocardia foam
- **d)** High return activated sludge concentration

PROBLEM 20.14

The activated sludge process can be designed and operated for removal of which of the following?

- **a)** Nonbiodegradable biochemical oxygen demand (BOD) and carbonaceous oxygen demand (COD)
- **b)** Ammonia and phosphorus
- **c)** Carbonaceous BOD, nitrogen, and phosphorus

PROBLEM 20.15

Typically, mixed liquor suspended solids (MLSS) concentrations fall within which of the following ranges?

    a)   1500 to 4500 mg/L
    b)   1000 to 3000 mg/L
    c)   950 to 2000 mg/L
    d)   2000 to 5000 mg/L

PROBLEM 20.16

Which of the following describes the meaning of F:M?

    a)   This is a measure of free oxygen in aeration tanks available to microorganisms.
    b)   This is a measure of free ammonia available to microorganisms.
    c)   This is a measure of how much food is available to microorganisms.
    d)   This is a measure of free carbon available to microorganisms.

PROBLEM 20.17

The volume of an aeration tank is 5650 m³ (1 500 000 gal). The MLSS is 2300 mg/L. If the aeration tank receives a primary effluent flow of 11 900 m³/d (3 150 000 gpd), with a primary effluent BOD of 110 mg/L, what is the food-to-microorganism ratio (F:M)?

    a)   0.10/d
    b)   0.01/d
    c)   0.03/d
    d)   0.08/d

PROBLEM 20.18

The flow to an aeration tank is 11 939 m³ (3 154 000 gpd), with a BOD of 172 mg/L. If the aeration tank is 36.6-m × 21.2-m × 4.9-m (120-ft × 70-ft × 16-ft) deep, the MLSS is 80% volatile, and the desired F:M is 0.25, what is the desired mixed liquor volatile suspended solids (MLVSS) concentration in the aeration tank?

    a)   3140 mg/L
    b)   2700 mg/L
    c)   2160 mg/L
    d)   1730 mg/L

PROBLEM 20.19

What is the current surface overflow rate (SOR) and solids loading rate (SLR), respectively, given the following information?

Three secondary clarifiers:

Parameter	Value
Diameter	40 m (130 ft)
Sidewall depth	4 m (13 ft)
Center depth	5.5 m (18 ft)
Return activated sludge	9000 mg/L
Return rate	40%
MLSS	3650 mg/L
Waste activated sludge rate	738 m³/d (0.195 mgd)
Flow	107 203 m³/d (28.32 mgd)

a)   90 m/d (2135 gpd/sq ft) and 469 kg/m²·d (97 lb/d/sq ft), respectively
b)   28 m/d (711 gpd/sq ft) and 145 kg/m2·d (30 lb/d/sq ft), respectively
c)   76 m/d (1809 gpd/sq ft) and 261 kg/m2·d (54 lb/d/sq ft), respectively
d)   25 m/d (603 gpd/sq ft) and 92 kg/m²·d (19 lb/d/sq ft), respectively

PROBLEM 20.20

Carbonaceous BOD measures which of the following?

a)   The oxygen demand associated with a combination of carbonaceous and nitrogenous biodegradable organic matter
b)   The oxygen demand associated with calcium-degrading organic matter
c)   The oxygen demand associated with carbonaceous biodegradable organic matter
d)   The oxygen demand associated with ammonia and nitrogen organic matter

PROBLEM 20.21

Mass loading rates in the activated sludge process can be expressed as which of the following?

a)   F:M
b)   Mean cell residence time and sludge volume index
c)   Sludge volume index and SRT
d)   Hydraulic retention time and COD

PROBLEM 20.22

Conventional activated sludge processes can remove what percentage of BOD and total suspended solids (TSS)?

a) 85 to 95%
b) 75 to 85%
c) 95 to 100%
d) 65 to 80%

PROBLEM 20.23

Calculate the mass of solids (kg/d) (lb/d) to be wasted every day to maintain a target SRT of 12 days, given the following information:

Parameter	Value
Flow	45 425 m³/d (12 mgd)
Effluent TSS	10 mg/L
MLSS	2850 mg/L
Bioreactor volume	18 549 m³ (4.9 mil. gal)
Waste activated sludge concentration	7500 mg/L

a) 4405 kg/d (9706 lb/d)
b) 3951 kg/d (8705 lb/d)
c) 3570 kg/d (7704 lb/d)
d) 2876 kg/d (6603 lb/d)

PROBLEM 20.24

Stalked ciliates are the predominant organisms seen when examining a sample of mixed liquor under a microscope. This indicates which of the following?

a) The mixed liquor is old and is in endogenous decay.
b) The mixed liquor is young and wasting rates should be reduced.
c) The mixed liquor is indicative of a nitrifying system.
d) The mixed liquor is healthy and the facility is running well.

Solutions

SOLUTION 20.1

The following are the basic system components of a conventional activated sludge process:

- Biological reactor,
- Secondary clarifier,
- Blowers and diffuser systems,
- Return and waste activated sludge removal, and
- Pumping systems.

SOLUTION 20.2

a) True

SOLUTION 20.3

a) Diagnosing and controlling bulking problems

SOLUTION 20.4

c) Organisms that use chemical compounds as their energy source

SOLUTION 20.5

b) Ammonia and nitrite

SOLUTION 20.6

Heterotrophs use <u>organic carbon</u> as their carbon source.

SOLUTION 20.7

c) Oxygen

SOLUTION 20.8

c) Aerobic

SOLUTION 20.9

Endogenous respiration

SOLUTION 20.10

The following are four environmental conditions that are important in the activated sludge process:

- pH,
- Temperature,
- Dissolved oxygen concentration, and
- Mixing intensity

SOLUTION 20.11

A selector creates environmental conditions that favor one type of organism over another.

SOLUTION 20.12

a) True

SOLUTION 20.13

a) Dispersed growth, pin floc, and bulking

SOLUTION 20.14

c) Carbonaceous BOD, nitrogen, and phosphorus

SOLUTION 20.15

b)   1000 to 3000 mg/L

SOLUTION 20.16

c)   This is a measure of how much food is available to microorganisms.

SOLUTION 20.17

a)   0.10/d

In International Standard units:

$$F:M = \frac{(BOD, mg/L) \times (Influent\ flow, m^3/d) \times \left[\dfrac{0.001\ kg/m^3}{mg/L}\right]}{(MLSS, mg/L) \times (Aerator\ volume, m^3) \times \left[\dfrac{0.001\ kg/m^3}{mg/L}\right]}$$

$$F:M = \frac{(110\ mg/L) \times (11\ 900, m^3/d) \times \left[\dfrac{0.001\ kg/m^3}{mg/L}\right]}{(2300\ mg/L) \times (5650\ m^3) \times \left[\dfrac{0.001\ kg/m^3}{mg/L}\right]} = 0.10\ d^{-1}$$

In U.S. customary units:

$$F:M = \frac{(BOD, mg/L) \times (Influent\ flow, mgd) \times \left[\dfrac{8.34\ lb/mil.\ gal}{mg/L}\right]}{(MLSS, mg/L) \times (Aerator\ volume, m^3) \times \left[\dfrac{8.34\ lb/mil.\ gal}{mg/L}\right]}$$

$$F:M = \frac{(110\ mg/L) \times (3.15\ mgd) \times \left[\dfrac{8.34\ lb/mil.\ gal}{mg/L}\right]}{(2300\ mg/L) \times (1.5\ mil.\ gal) \times \left[\dfrac{8.34\ lb/mil.\ gal}{mg/L}\right]} = 0.10\ d^{-1}$$

SOLUTION 20.18

c)   2160 mg/L

In International Standard units:

$$\text{MLVSS} = \frac{(\text{BOD, mg/L}) \times (\text{Influent flow, m}^3/\text{d})}{(\text{Aerator volume, m}^3) \times (\text{F:M, d}^{-1})}$$

$$\text{MLVSS} = \frac{(11\,939 \text{ m}^3/\text{d}) \times (172 \text{ mg/L})}{(36.6 \text{ m} \times 21.2 \text{ m} \times 4.9 \text{ m}) \times (0.25 \text{ d}^{-1})} = 2160 \text{ mg/L}$$

In U.S. customary units:

$$\text{MLVSS} = \frac{(\text{BOD, mg/L}) \times (\text{Influent flow, mgd})}{(\text{Aerator volume, cu ft})(7.48 \text{ gal/cu ft}) \times (\text{F:M, d}^{-1})}$$

$$\text{MLVSS} = \frac{(3\,154\,000 \text{ gpd}) \times (172 \text{ mg/L})}{(120 \text{ ft} \times 70 \text{ ft} \times 16 \text{ ft}) \times (7.48 \text{ gal/cu ft}) \times (0.25 \text{ d}^{-1})}$$

$$= 2158 \text{ mg/L (round to 2160 mg/L)}$$

SOLUTION 20.19

b)   28 m/d (711 gpd/sq ft) and 145 kg/m² ·d (30 lb/d/sq ft), respectively

In International Standard units:

$$\text{SOR} = Q/A$$

$$\text{SOR} = \frac{(107\,203 \text{ m}^3/\text{d})}{3.14(20 \text{ m}) \times 3} = 28 \text{ m/d}$$

$$\text{SLR} = [(Q + Q_r) \times \text{MLSS}]/A$$

$$\text{SLR} = \frac{(107\,203 \text{ m}^3/\text{d}) + (0.4 \times 107\,203 \text{ m}^3/\text{d}) \times (3650 \text{ mg/L}) \times \left(\frac{0.001 \text{ kg/m}^3}{\text{mg/L}}\right)}{3.14(20 \text{ m})^2 \times 3}$$

$$= 145 \text{ kg/m}^2 \cdot \text{d}$$

In U.S. customary units:

$$SOR = Q/A$$

$$SOR = \frac{28\,320\,000 \text{ gal}}{3.14(65 \text{ ft})^2 \times 3} = 711 \text{ gpd/sq ft}$$

$$SLR = [(Q + Q_r) \times MLSS]/A$$

$$SLR = \frac{(28.32 \text{ mgd}) + (0.4 \times 28.32 \text{ mgd}) \times \left(\dfrac{8.34 \text{ lb/mil. gal}}{\text{mg/L}}\right) \times (3650 \text{ mg/L})}{3.14(65 \text{ ft})^2 \times 3}$$

$$= 30/\text{lb/d/sq ft}$$

SOLUTION 20.20

c)   The oxygen demand associated with carbonaceous biodegradable organic matter

SOLUTION 20.21

a)   F:M

SOLUTION 20.22

a)   85 to 95%

SOLUTION 20.23

b)   3951 kg/d (8705 lb/d)

In International Standard units:

$$\text{Waste activated sludge} = \frac{(18\,549 \text{ m}^3) \times (2850 \text{ mg/L}) \times \left(\dfrac{0.001 \text{ kg/m}^3}{\text{mg/L}}\right)}{12 \text{ days}}$$

$$- (45\,425 \text{ m}^3/\text{d}) \times (10 \text{ mg/L})\left(\dfrac{0.001 \text{ kg/m}^3}{\text{mg/L}}\right) = 3951 \text{ kg/d}$$

In U.S. customary units:

$$\text{Waste activated sludge} = \frac{(4.9 \text{ mil. gal}) \times (2850 \text{ mg/L}) \times \left(\dfrac{8.34 \text{ lb/mil. gal}}{\text{mg/L}}\right)}{12 \text{ days}}$$

$$- (12 \text{ mgd}) \times (10 \text{ mg/L})\left(\frac{8.34 \text{ lb/mil. gal}}{\text{mg/L}}\right) = 8705 \text{ lb/d}$$

SOLUTION 20.24

d) The mixed liquor is healthy and the facility is running well.

Chapter 21
TRICKLING FILTERS, ROTATING BIOLOGICAL CONTACTORS, AND COMBINED PROCESSES

Problems

PROBLEM 21.1

Trickling filters and rotating biological contactors (RBCs) are fixed-film processes. What is a fixed film?

  a) A thin sheet of plastic covering the biological treatment system to protect it from algal growth
  b) A viscous, jellylike slime with living organisms, including bacteria, protozoa, algae, and fungi
  c) A jellylike film made up of glucose, organic polymers, and worms

PROBLEM 21.2

Fixed-film systems are affected by variations in which of the following?

  a) pH, alkalinity, biochemical oxygen demand (BOD), and temperature
  b) pH, total suspended solids, and grit
  c) Alkalinity, total Kjeldahl nitrogen, total nitrogen, and nitrate

PROBLEM 21.3

What are the organic and hydraulic loading, respectively, to one trickling filter with a diameter of 36 m (120 ft) and depth of 2 m (7 ft), Q = 17 034 m³/d (4.5 mgd), recirculation flow of 50%, primary effluent biochemical oxygen demand (BOD) of 140 mg/L, and total pumping capacity of 343 608 m³/d (8000 gpm)?

a) 1.17 kg BOD/m³·d (0.0066 lb BOD/d/cu ft) and 43 m/d (0.71 gpm/sq ft), respectively

b) 17 kg BOD/100 m³·d (66 lb BOD/d/1000 cu ft) and 43 m/d (0.71 gpm/sq ft), respectively

c) 117 kg BOD/100 m³·d (66 lb BOD/d/1000 cu ft) and 43 m/d (0.71 gpm/sq/ft), respectively

PROBLEM 21.4

The typical organic loading to a trickling filter with rock media is 90 kg BOD/100 m³·d (50 lb BOD/d/1000 cu ft).

a) True

b) False

PROBLEM 21.5

List four purposes served by recirculation in a trickling filter process.

PROBLEM 21.6

Calculate the total and soluble organic and hydraulic loads on an RBC system, given the following Q = 15 900 m³/d (4.2 mgd), primary effluent BOD = 115 mg/L, primary effluent soluble BOD (sBOD) = 92 mg/L, and total number of RBC shafts = 9 at 9300 m²/shaft (100 000 sq ft/shaft).

a) 21.8 g BOD₅/m²·d (4.48 lb BOD₅/d/1000 sq ft), 17.5 g sBOD₅/m²·d (3.58 lb sBOD₅/d/1000 sq ft), and 0.19 m³/m²·d (4.67 gpd/sq ft), respectively

b) 218 g BOD₅/m²·d (4480 lb BOD₅/d/1000 sq ft), 175 g sBOD₅/m²·d (3580 lb sBOD₅/d/1000 sq ft), and 0.19 m³/m²·d (4.67 gpd/sq ft), respectively

c) 2.18 g BOD₅/m²·d (44.8 lb BOD₅/d/1000 sq ft), 1.75 g sBOD₅/m²·d (35.8 lb sBOD₅/d/1000 sq ft), and 0.19 m³/m²·d (4.67 gpd/sq ft), respectively

PROBLEM 21.7

Rotating biological contactors and trickling filters are different from activated sludge in that they do not require which of the following?

a)   Oxygen
b)   Nutrients
c)   Returning sludge

PROBLEM 21.8

Reddish-brown patches on an RBC is an indication of Beggiatoa and Thiothrix.

a)   True
b)   False

PROBLEM 21.9

For nitrification, high-density media are typically used in RBCs. High-density media per shaft are considered to be which of the following?

a)   10 300 m² (112 000 sq ft)
b)   11 200 m² (120 000 sq ft)
c)   18 300 m² (197 000 sq ft)

PROBLEM 21.10

What are two reasons for excessive sloughing from an RBC?

PROBLEM 21.11

Typical BOD loading to an RBC would be which of the following?

a)   1.2 to 2.0 kg sBOD/100 m²·d (2.5 to 4 lb sBOD/d/1000 sq ft)
b)   3.0 to 4.0 kg sBOD/100 m²·d (1.5 to 2.0 lb sBOD/d/1000 sq ft)
c)   1.5 to 2.0 kg sBOD/m²·d (3 to 4 lb sBOD/d/sq ft)

PROBLEM 21.12

Typical hydraulic loading to an RBC for carbonaceous BOD removal would be which of the following?

   a)   0.04 to 0.12 $m^3/m^2 \cdot d$ (1.0 to 3.0 gpd/sq ft)
   b)   0.004 to 0.012 $m^3/m^2 \cdot d$ (1.0 to 3.0 gpd/sq ft)
   c)   0.4 to 1.2 $m^3/m^2 \cdot d$ (10 to 30 gpd/sq ft)

PROBLEM 21.13

What is a combined process?

   a)   BOD and nitrogen removal
   b)   Activated sludge followed by an RBC
   c)   Trickling filter followed by activated sludge

PROBLEM 21.14

The advantage of a combined system is which of the following?

   a)   Lower polymer costs
   b)   Reduced influent BOD to the activated sludge process
   c)   Removal of both phosphorus and nitrogen

PROBLEM 21.15

A trickling filter placed before an activated sludge system is sometimes called which of the following?

   a)   Roughing filter
   b)   Flow attenuating filter
   c)   Polishing filter

Solutions

SOLUTION 21.1

   b)   A viscous, jellylike slime with living organisms, including bacteria, protozoa, algae, and fungi

SOLUTION 21.2

a)   pH, alkalinity, BOD, and temperature

SOLUTION 21.3

b)   117 kg BOD/100 m³·d (66 lb BOD/d/1000 cu ft) and 43 m/d (0.71 gpm/sq ft), respectively

In International Standard units:

$$\text{Filter area} = 3.14(18 \text{ m})^2 = 1017 \text{ m}^2$$

$$\text{Filter volume} = 1017 \text{ m}^2 \times 2 \text{ m} = 2034 \text{ m}^3/100 = 20.34 \, (100 \text{ m}^3)$$

$$\text{Organic loading} = 140 \text{ mg/L} \times 17\,032 \text{ m}^3/\text{d} \times \left(\frac{0.001 \text{ kg/m}^3}{\text{mg/L}} \right) = 2384 \text{ kg BOD/d}$$

$$\frac{2384 \text{ kg BOD/d}}{20.34 \, (100 \text{ m}^3)} = 117 \text{ kg BOD/100 m}^3 \cdot \text{d}$$

$$\text{Hydraulic loading} = \frac{(43\,608 \text{ m}^3/\text{d})}{(1017 \text{ m}^2)} = 43. \text{ m}^3/\text{m}^2 \cdot \text{d} = 43 \text{ m/d}$$

In U.S. customary units:

$$\text{Filter area} = 3.14(60 \text{ ft})^2 = 11\,304 \text{ sq ft}$$

$$\text{Filter volume} = 1\,304 \text{ sq ft} \times 7 \text{ ft} = 79\,128 \text{ cu ft}/1000 = 79.13 \, (1000 \text{ cu ft})$$

$$\text{Organic loading} = 140 \text{ mg/L} \times \left(\frac{8.34 \text{ lb/mil. gal}}{\text{mg/L}} \right) \times 4.5 \text{ mgd} = 5254 \text{ lb BOD/d}$$

$$\frac{5254 \text{ lb BOD/d}}{79.13 \, (1000 \text{ cu ft})} = 66 \text{ lb BOD/d } 1000 \text{ cu ft}$$

$$\text{Hydraulic loading} = 8000 \text{ gpm}/11\,304 \text{ sq ft} = 0.71 \text{ gpm/sq/ft}$$

SOLUTION 21.4

a)   True

SOLUTION 21.5

Four purposes served by recirculation in a trickling filter process include any four of the following:

- Reducing the strength of wastewater applied to the filter;
- Increasing the hydraulic load to reduce flies, snails, and other nuisances;
- Maintaining distributor movement during low flows;
- Diluting toxic wastes;
- Reseeding the filter's microbial population;
- Providing uniform distribution of flow;
- Preventing filters from drying out; and
- Producing hydraulic shear to encourage solids sloughing and prevent ponding.

SOLUTION 21.6

a) 21.8 g BOD$_5$/m²·d (4.48 lb BOD$_5$/d/1000 sq ft), 17.5 g sBOD$_5$/m²·d (3.58 lb sBOD$_5$/d/1000 sq ft), and 0.19 m³/m²·d (4.67 gpd/sq ft), respectively

In International Standard units:

$$\text{Organic loading} = \frac{(15\,900\ \text{m}^3/\text{d}) \times (115\ \text{mg/L}) \times \left(\dfrac{1000\ \text{L/m}^3}{1000\ \text{mg/g}}\right)}{9 \times 9300\ \text{m}^2} = 21.8\ \text{g BOD}_5/\text{m}^2 \cdot \text{d}$$

$$\text{Soluble Organic loading} = \frac{(15\,900\ \text{m}^3/\text{d}) \times (92\ \text{mg/L}) \times \left(\dfrac{1000\ \text{L/m}^3}{1000\ \text{mg/g}}\right)}{9 \times 9300\ \text{m}^2} = 17.5\ \text{g BOD}_5/\text{m}^2 \cdot \text{d}$$

$$\text{Hydraulic loading} = \frac{(15\,900\ \text{m}^3/\text{d})}{(9 \times 9300\ \text{m}^2)} = 0.19\ \text{m}^3/\text{m}^2 \cdot \text{d}$$

In U.S. customary units:

$$\text{Organic loading} = \frac{4.2 \text{ mgd} \times \left(\dfrac{8.34 \text{ lb/mil. gal}}{\text{mg/L}} \right) \times 115 \text{ mg/L}}{\dfrac{9 \times 100\,000 \text{ sq ft}}{(1000)}} = 4.48 \text{ lb BOD}_5/\text{d}/1000 \text{ sq ft}$$

$$\text{Soluble Organic loading} = \frac{4.2 \text{ mgd} \times \left(\dfrac{8.34 \text{ lb/mil. gal}}{\text{mg/L}} \right) \times 92 \text{ mg/L}}{\dfrac{9 \times 100\,000 \text{ sq ft}}{(1000)}} \times 3.58 \text{ lb sBOD}_5/\text{d}/1000 \text{ sq ft}$$

$$\text{Hydraulic loading} = \frac{4\,200\,000 \text{ gpd}}{9 \times 100\,000 \text{ sq ft}} = 4.67 \text{ gpd/sq/ft}$$

SOLUTION 21.7

c)   Returning sludge

SOLUTION 21.8

b)   False

SOLUTION 21.9

b)   11 200 m² (120 000 sq ft)

SOLUTION 21.10

The following are two reasons for excessive sloughing from an RBC:

- Toxic material in the influent and
- Excessive pH variations (below 5 and above 10).

SOLUTION 21.11

a) 1.2 to 2.0 kg sBOD/100 m²·d (2.5 to 4 lb sBOD/d/1000 sq ft)

SOLUTION 21.12

a) 0.04 to 0.12 m³/m²·d (1.0 to 3.0 gpd/sq ft)

SOLUTION 21.13

c) Trickling filter followed by activated sludge

SOLUTION 21.14

b) Reduced influent BOD to the activated sludge process

SOLUTION 21.15

a) Roughing filter

Chapter 22
BIOLOGICAL NUTRIENT REMOVAL PROCESSES

Problems

PROBLEM 22.1

A water resource recovery facility that is nitrifying will use how much more oxygen than just one removing carbonaceous biochemical oxygen demand?

- **a)** 20 to 30%
- **b)** 10 to 20%
- **c)** 50 to 60%
- **d)** 30 to 40%

PROBLEM 22.2

Nitrification requires a pH range of _____ and a dissolved oxygen range of _____.

- **a)** 6.5 to 8.0 and 2 to 3 mg/L, respectively
- **b)** 7.2 to 7.5 and 0.5 to 1.0 mg/L, respectively
- **c)** 7.5 to 9.0 and 3 to 5 mg/L, respectively
- **d)** 6.5 to 8.0 and no dissolved oxygen, respectively

PROBLEM 22.3

Nitrification uses alkalinity because of the production of mineral acids. These acids include which of the following?

a)   Hydrochloric and hydrochlorous acid

b)   Phosphoric and phosphorus acid

c)   Nitrous and nitric acid

d)   Hypochlorite and hypochlorous acid

PROBLEM 22.4

How much alkalinity (as calcium carbonate, $CaCO_3$) is returned by denitrification of 20 mg/L of nitrate-nitrogen (NO_3-N)?

a)   None

b)   71.4 mg/L

c)   152 mg/L

d)   7.2 mg/L

PROBLEM 22.5

Phosphorus-accumulating organisms (PAOs) use readily available substrate in the anaerobic zone for which purpose?

a)   Growth

b)   To gain energy

c)   To convert to internal storage products

d)   To reduce phosphorus

PROBLEM 22.6

In a combined biological nutrient removal (BNR) system, why do we care about aerobic solids retention time (SRT)?

a)   Because SRT is related to the nitrifying fraction and nitrifying performance

b)   Because SRT controls eutrophication

c)   Because SRT reduces the phosphorus-accumulating organism population

d)   Because SRT is related to the denitrification capacity of the system

PROBLEM 22.7

Denitrification is an oxidation process to remove nitrogen.

a) True
b) False

PROBLEM 22.8

In the luxury uptake of the phosphorus process, under what conditions do microorganisms release phosphorus from their cells?

a) Activated sludge
b) Aerobic
c) Anaerobic
d) Clarification

PROBLEM 22.9

Methanol is used in the BNR process to do which of the following?

a) Reduce foaming
b) Supplement carbon for efficient nitrification
c) Supplement carbon for efficient denitrification
d) Increase the chemical oxygen demand to total Kjeldahl nitrogen ratio

PROBLEM 22.10

If a wastewater environment has no dissolved oxygen but has nitrate, it is described as which of the following?

a) Aerobic
b) Anaerobic
c) Anoxic
d) Hypoxic

PROBLEM 22.11

The denitrification process "gives back" _____ mg/L of alkalinity and _____ mg/L of oxygen equivalence.

a) 7.14 and 4.6, respectively
b) 3.57 and 2.86, respectively
c) 4.6 and 7.14, respectively
d) 2.86 and 3.6, respectively

PROBLEM 22.12

Biomass in a conventional activated sludge process contains approximately what percentage of phosphorus?

a) 1%
b) 3%
c) 4%
d) 2%

PROBLEM 22.13

In an activated sludge facility designed for phosphorus removal, biomass can contain what percentage of phosphorus?

a) 1 to 4%
b) 2 to 5%
c) 3 to 6%
d) 4 to 7%

PROBLEM 22.14

In the luxury uptake of phosphorus, the operator must carefully regulate the detention time so that it is long enough to remove as much phosphorus as possible, but not so long that the organisms die.

a) True
b) False

PROBLEM 22.15

Bio-P processes can produce effluent with how much total phosphorus?

  a)   1 to 2 mg/L
  b)   2 to 3 mg/L
  c)   3 to 4 mg/L
  d)   4 to 5 mg/L

PROBLEM 22.16

Phosphorus is typically removed in the treatment process by one or two distinct processes. To achieve very low concentrations of phosphorus in the effluent, these two processes are often combined. Name these two processes.

PROBLEM 22.17

Calculate how much alkalinity would be required for a nitrification process if you want 65 mg/L of alkalinity left in the effluent. The influent ammonia concentration is 17.8 mg/L and you get 98% conversion of ammonia. The flow at the facility is 13 627 m^3/d (3.6 mgd).

  a)   811 kg/d (1760 lb/d)
  b)   2589 kg/d (5705 lb/d)
  c)   4618 kg/d (7567 lb/d)
  d)   1762 kg/d (3804 lb/d)

PROBLEM 22.18

If a facility has the following configuration in the biological reactors, what compounds would be removed within the following process: anaerobic zone, followed by anoxic zone, followed by an aerobic zone.

  a)   Carbon, nitrogen, and alkalinity
  b)   Phosphorus, nitrogen, and alkalinity
  c)   Carbon, nitrogen, and volatile acids
  d)   Carbon, nitrogen, and phosphorus

PROBLEM 22.19

The chlorine demand at your facility suddenly increases significantly. This change is most likely caused by what?

   a)   Nitrification beginning to occur at the facility
   b)   An industrial discharge upsetting the facility's pH balance
   c)   A sudden drop in ambient temperatures
   d)   Excessive mixing in the chlorine contact chamber

PROBLEM 22.20

The manager of a facility is asked to determine alkalinity requirements for the new nitrification/denitrification process. Determine whether alkalinity will need to be added and, if so, how much it will cost per year, given the following:

Influent ammonia = 27.9 mg/L
Primary effluent ammonia = 29.2 mg/L
Final effluent ammonia = 1.2 mg/L
Influent nitrate = 0
Primary effluent nitrate = 0
Final effluent nitrate = 4.9 mg/L
Influent flow = 54 504 m³/d (14.4 mgd)
Primary effluent alkalinity = 134 mg/L
Desired final effluent alkalinity = 50 mg/L
Cost of sodium carbonate ($NaCO_3$) = $0.087/kg ($0.0395/lb)

Solutions

SOLUTION 22.1

   d)   30 to 40%

SOLUTION 22.2

   a)   6.5 to 8.0 and 2 to 3 mg/L, respectively

SOLUTION 22.3

   c)   Nitrous and nitric acid

SOLUTION 22.4

b) 71.4 mg/L (20 mg/L NO_3-N) × (3.57 mg $CaCO_3$/mg NO_3-N removed)

SOLUTION 22.5

c) To convert to internal storage products

SOLUTION 22.6

a) Because it is related to the nitrifying fraction and nitrifying performance

SOLUTION 22.7

b) False

SOLUTION 22.8

c) Anaerobic

SOLUTION 22.9

c) Supplement carbon for efficient denitrification

SOLUTION 22.10

c) Anoxic

SOLUTION 22.11

b) 3.57 and 2.86, respectively

SOLUTION 22.12

d) 2%

SOLUTION 22.13

c)   3 to 6%

SOLUTION 22.14

a)   True

SOLUTION 22.15

a)   1 to 2 mg/L

SOLUTION 22.16

Biological phosphorus removal and chemical precipitation

SOLUTION 22.17

b)   2589 kg/d (5705 lb/d)

Alkalinity needed = Alkalinity required + Alkalinity goal

Alkalinity required = [(17.8 mg/L NH_3-N*) $\times$ 0.98 $\times$ (7.14 mg $CaCO_3$/mg NH_3-N)]
+ 65 mg/L = 190 mg/L

In International Standard units:

$$\text{Alkalinity required} = (190 \text{ mg/L}) \times (13\,627 \text{ m}^3/\text{d}) \times \left(\frac{0.001 \text{ kg/m}^3}{\text{mg/L}} \right) = 2589 \text{ kg/d}$$

*NH_3-N = ammonia-nitrogen

In U.S. customary units:

$$\text{Alkalinity required} = (190 \text{ mg/L}) \times (3.6 \text{ mgd}) \times \left(\frac{8.34 \text{ lb/mil. gal}}{\text{mg/L}} \right) = 5705 \text{ lb/d}$$

SOLUTION 22.18

d)   Carbon, nitrogen, and phosphorus

SOLUTION 22.19

a)   Nitrification beginning to occur at the facility

SOLUTION 22.20

$58 900 per year

$$NH_3\text{-}N \text{ converted to } NO_3\text{-}N = 29.2 \text{ mg/L} - 1.2 \text{ mg/L} = 28 \text{ mg/L}$$

$$NO_3\text{-}N \text{ removed} = 27.9 \text{ mg/L} - 4.9 \text{ mg/L} = 23 \text{ mg/L}$$

$$\text{Alkalinity needed for nitrification} = 28 \text{ mg/L} \times (7.14 \text{ mg CaCO}_3/\text{mg NH}_3\text{-}N)] = 200 \text{ mg/L}$$

$$\text{Alkalinity returned by denitrification} = 23 \text{ mg/L} \times 3.57 \text{ mg CaCO}_3/\text{mg NH}_3\text{-}N) = 82 \text{ mg/L}$$

$$\text{Alkalinity required} = 200 \text{ mg/L} - 82 \text{ mg/L} + 50 \text{ mg/L} = 168 \text{ mg/L}$$

$$\text{Alkalinity in primary effluent} = 134 \text{ mg/L}$$

$$\text{Alkalinity to be added per day} = 168 \text{ mg/L} - 134 \text{ mg/L} = 34 \text{ mg/L}$$

In International Standard units:

$$\text{Alkalinity to be added} = (34 \text{ mg/L}) \times (54\,504 \text{ m}^3/\text{d}) \times \left(\frac{0.001 \text{ kg/m}^3}{\text{mg/L}} \right) = 1853 \text{ kg/d}$$

$$\text{Cost} = 1853 \text{ kg/d} \times \$0.087/\text{kg} \times 365 \text{ d/yr} = \$58\,842/\text{yr} \text{ (round to } \$58\,900/\text{yr)}$$

In U.S. customary units:

$$\text{Alkalinity to be added} = (34 \text{ mg/L}) \times (14.4 \text{ mgd}) \times \left(\frac{8.34 \text{ lb/mil. gal}}{\text{mg/L}} \right) = 4083 \text{ lb/d}$$

$$\text{Cost} = 4083 \text{ lb/d} \times \$0.0395/\text{lb} \times 365 \text{ d/yr} = \$58\,866/\text{yr} \text{ (round to } \$58\,900/\text{yr)}$$

Chapter 23
NATURAL BIOLOGICAL PROCESSES

Problems

PROBLEM 23.1

Nitrogen is removed in a land treatment system by all but which of the following?

- **a)** Ammonia volatilization
- **b)** Nitrification
- **c)** Plant uptake
- **d)** Denitrification

PROBLEM 23.2

Phosphorus is primarily removed from wastewater in a slow-rate land treatment system by which of the following?

- **a)** Soil adsorption and chemical precipitation
- **b)** Surface sedimentation
- **c)** Vegetative adsorption
- **d)** Bacterial assimilation by soil microorganisms

PROBLEM 23.3

All land treatment systems are required to have a National Pollutant Discharge Elimination System (NPDES) permit.

- **a)** True
- **b)** False

PROBLEM 23.4

The following part of the plant will be first to show any increase in metal content, thus providing an early indication of problems with the slow-rate land treatment process.

a) Roots
b) Mature stalks
c) Leaves
d) Fruit/seeds

PROBLEM 23.5

In cold climates, the annual disking of the surface soil in a rapid infiltration bed will do which of the following?

a) Aid in the volatilization of ammonia
b) Increase phosphorus adsorption
c) Keep the beds from clogging during the winter
d) Prevent the soil from freezing

PROBLEM 23.6

A facultative treatment lagoon for a cold climate is to be designed for a community having an average daily flow of 567 m^3/d (0.15 mgd) and an average daily influent 5-day biochemical oxygen demand (BOD$_5$) of 210 mg/L. The regulatory requirements are as follows: the organic loading is not to exceed 22 kg/ha·d (19.6 lb/d/ac), the bottom 0.6 m (2 ft) is to be reserved for residuals storage and is not to be drained, the overall liquid depth is not to exceed 1.85 m (6 ft), and the lagoon is to have a semiannual discharge. What is the minimum required surface area for the lagoon? (For simplicity, assume vertical side slopes.)

a) 36 686 m^2 (394 112 sq ft)
b) 54 130 m^2 (582 000 sq ft)
c) 55 200 m^2 (594 009 sq ft)
d) 81 648 m^2 (878 532 sq ft)

PROBLEM 23.7

Freeboard is provided for a lagoon for which of the following reasons?

a) To provide extra capacity for rainfall events
b) For dike stability and room for ice accumulation
c) To protect inlet structure and prevent short-circuiting
d) To permit lowering of water level and control depth

PROBLEM 23.8

Which of the following materials may be used to prevent seepage from a lagoon?

a) Bentonite clay
b) Chlorosulfonated polyethylene
c) Vinyl
d) All of the above

PROBLEM 23.9

In land-based treatment systems such as irrigation, how often should soil testing be conducted?

a) Quarterly
b) Monthly
c) Annually
d) Weekly

PROBLEM 23.10

Which of the following conditions may create a pH of below 6.5 and odor problems in a lagoon?

a) Water temperatures higher than 28 °C (82 °F) during the day and the appearance of significant algae growth in the lagoon
b) Scum and weed growth covering the surface and breeding insects
c) Low dissolved oxygen, creating anaerobic conditions favorable for acid bacteria
d) High levels of oil and grease in the influent, promoting the growth of nuisance bacteria and blue–green algae

PROBLEM 23.11

Surface aerators may be installed in a lagoon to produce better effluent when

    a)   The lagoon is hydraulically overloaded.
    b)   High levels of BOD are carried in the influent.
    c)   The influent contains a high level of nonvolatile suspended solids.
    d)   The lagoon is covered by algal mats.

PROBLEM 23.12

To control odors from a lagoon, how much sodium nitrate must be added initially and on succeeding days to a 2.33-ha (5.75-ac) lagoon?

    a)   136 kg (300 lb) and 79 kg (175 lb), respectively
    b)   260 kg (575 lb) and 130 kg (287.5 lb), respectively
    c)   501 kg (1150 lb) and 196 kg (431.5 lb), respectively
    d)   714 kg (1575 lb) and 238 kg (525 lb), respectively

PROBLEM 23.13

Aerobic and facultative lagoons appear _____ in color as a result of higher productivity of primary cells in summer, whereas they appear _____/_____ during winter.

PROBLEM 23.14

Surface runoff is an important parameter for designing a slow-rate land treatment process.

    a)   True
    b)   False

PROBLEM 23.15

What is a typical concentration of dissolved oxygen in a properly operated aerated lagoon?

    a)   1 to 2 mg/L
    b)   3 to 4 mg/L
    c)   5 to 6 mg/L
    d)   7 to 8 mg/L

PROBLEM 23.16

The presence of large amounts of blue–green algae in lagoons is an indication of which of the following?

a)   Poor nutrient balance
b)   Overloading
c)   Incomplete treatment
d)   All of the above

PROBLEM 23.17

In most situations, what is the primary factor limiting the application of wastes to the soil?

a)   Phosphorus
b)   Nitrogen
c)   Toxic substances
d)   Sodium adsorption ratio

PROBLEM 23.18

The dark gray to black color for aerobic lagoons indicates which of the following?

a)   Daphnia blooms
b)   Algae blooms
c)   Anaerobic conditions
d)   Presence of excess dissolved oxygen

PROBLEM 23.19

The best growing conditions for most crops occur when the moisture in the soil is at which of the following?

a)   Saturation
b)   Zero
c)   Field capacity
d)   One-half of the field capacity

PROBLEM 23.20

Calculate the maximum wastewater flowrate that can be treated with the land treatment system described below. Assume that 20% of applied nitrogen is lost as a result of denitrification and volatilization. Use the following data:

Land surface area designed for disposal of wastewater = 123.5 ha (305.2 ac)

Precipitation rate = 0.12 m/mo (4.7 in./mo)

Evapotranspiration rate = 0.07 m/mo (2.75 in./mo)

Percolation rate = 0.25 m/mo (9.84 in./mo)

Initial nitrogen concentration in applied wastewater (C_n) = 40 mg/L

Percolate nitrogen concentration (i_p) = 10 mg/L

Nitrogen uptake by crop = 200 kg/ha·yr

Solutions

SOLUTION 23.1

b)   Nitrification

Nitrogen is removed through crop uptake and harvest. Denitrification can be a mechanism for nitrogen loss as well as ammonia volatilization and storage in the soil. Nitrification is the conversion of ammonia to nitrate. However, the nitrogen is still present and has not been removed.

SOLUTION 23.2

a)   Soil adsorption and chemical precipitation

SOLUTION 23.3

b)   False

All land treatment systems with point-source discharges have an NPDES permit that establishes performance and operational criteria for the system. Land application systems without point-source discharges may not necessarily be permitted or have a groundwater permit.

SOLUTION 23.4

c)   Leaves

Typically, the leafy material will be the first to show increases in metal content, thereby providing an early indication of future problems.

SOLUTION 23.5

c)   Keep the beds from clogging during the winter

In cold climates, the surface soil should be disked each year during the late summer or fall to keep the beds from clogging during the winter.

SOLUTION 23.6

d)   81 648 m² (878 532 sq ft)

Determine required volume based on loading:

$$\text{Organic loading} = (567 \text{ m}^3/\text{d}) \times (210 \text{ mg/L}) \times \left(\frac{0.001 \text{ kg/m}^3}{\text{mg/L}} \right) = 119 \text{ kg/d BOD (262 lb/d BOD)}$$

$$\text{Surface area required} = \frac{119 \text{ kg/d}}{22 \text{ kg/ha} \cdot \text{d}} = 5.41 \text{ ha (13.4 ac)}$$

$$\text{Maximum storage depth} = 1.85 \text{ m} - 0.6 \text{ m} = 1.25 \text{ m (4 ft)}$$

$$\text{Resultant storage volume} = 5.41 \text{ ha} \times 10\,000 \text{ m}^2/\text{ha} \times 1.25 \text{ m} = 67\,625 \text{ m}^3 \text{ (17.87 mil. gal)}$$

Determine required volume based on a semiannual discharge:

$$\text{Minimum days for storage} = 180 \text{ days}$$

$$\text{Resultant storage required} = 180 \text{ d} \times 567 \text{ m}^3/\text{d} = 102\,060 \text{ m}^3 \text{ (26.96 mil. gal)}$$

Volume based on the storage requirement governs:

$$\text{Required surface area} = 102\,060 \text{ m}^3/1.25 \text{ m} = 81\,648 \text{ m}^2 \text{ (878 532 sq ft)}$$

SOLUTION 23.7

b)   For dike stability and room for ice accumulation

Larger lagoons include a minimum of 0.6 to 0.9 m (2 to 3 ft) of freeboard that provides stability to the dike and room for ice accumulation in colder climates.

SOLUTION 23.8

d)   All of the above

To prevent seepage from the lagoon, a natural clay liner is typically provided. Bentonite and vinyl liners have also been used to limit seepage.

Some liner materials on the market today, such as chlorosulfonated polyethylene, are resistant to UV light damage.

SOLUTION 23.9

c)   Annually

Soil samples should be collected from each irrigated area at least annually.

SOLUTION 23.10

c)   Low dissolved oxygen, creating anaerobic conditions favorable for acid bacteria

Low pH in anaerobic ponds; a pH below 6.5 accompanied by odors results from acid bacteria working in the anaerobic condition.

SOLUTION 23.11

b)   High levels of BOD are carried in the influent

Poor light penetration, low detention time, high biochemical oxygen demand (BOD), loading or toxic wastes (daytime dissolved oxygen should not drop below 3.0 mg/L during warm months) may cause a low, continued downward trend in dissolved oxygen. Solution: Add supplemental aeration (surface aerators, diffusers, or daily operation of a motorboat).

SOLUTION 23.12

b)   260 kg (575 lb) and 53 kg (287.5 lb), respectively

In International Standard units:

$$\text{Initially, } 112 \text{ kg/ha} \times 2.33 \text{ ha} = 260 \text{ kg}$$

$$\text{On succeeding days, } 56 \text{ kg/ha} \times 2.33 \text{ ha} = 130 \text{ kg}$$

In U.S. customary units:

Initially, 100 lb/ac × 5.75 ac = 575 lb

On succeeding days, 50 lb/ac × 5.75 ac = 287.5 lb

Apply chemicals such as sodium nitrate, or 1,2-dibromo-2,2-dichloroethyl dimethylphos-phate, to introduce oxygen; application rate: 5 to 15% sodium nitrate/mil. gal [mil. gal × (3.785 × 103) = m³]; repeat at a reduced rate on succeeding days; use 100 lb sodium ni-trate/ac (112 kg/ha) for first day, then 50 lb/ac/day (56 kg/ha·d) thereafter if odors persist; apply in the wake of a motorboat.

SOLUTION 23.13

Aerobic and facultative lagoons appear <u>green</u> in color as a result of higher productivity of primary cells in summer, whereas they appear <u>brown/gray</u> during winter.

During the warmer months, primary cells should be highly productive and have a green color, indicating high pH and dissolved oxygen.

In the late fall and winter, biological activity will decrease and the color of the lagoons will change to brown and then gray.

SOLUTION 23.14

b)   False

Slow-rate systems are designed for no surface runoff.

SOLUTION 23.15

a)   1 to 2 mg/L

Typical practice is to keep 1 to 2 mg/L of dissolved oxygen in the lagoon.

SOLUTION 23.16

d)   All of the above

Blue–green algae is an indication of incomplete treatment, overloading, or poor nutrient balance.

SOLUTION 23.17

b) Nitrogen

In most situations, nitrogen is the main factor limiting the application of waste to the soil.

SOLUTION 23.18

c) Anaerobic conditions

Gray to black color in the lagoon is undesirable.

SOLUTION 23.19

c) Field capacity

The best growing conditions for most crops occur when the moisture in the soil is at field capacity. Field capacity is the soil moisture content (percent moisture on a dry weight basis) in the field 2 or 3 days after saturation and after free drainage has ceased, that is, the quantity of water held in the soil by capillary action after gravitational or free water has drained from the soil.

SOLUTION 23.20

4000 m³/d (733.9 gpm)

Assume that X m³/d of wastewater flow can be treated with the given land treatment system. Hydraulic loading rate (LA) based on water balance is determined as follows:

$$PR + LA = ET + SP$$

Where

        PR = precipitation rate,
        ET = evapotranspiration rate; and
        SP = soil percolation rate.

$$\therefore LA = ET + SP \, \text{‴} \, PR$$

$$\therefore LA = 0.07 \text{ m/mo} + 0.25 \text{ m/mo} - 0.12 \text{ m/mo}$$

$$\therefore LA = 0.20 \text{ m/mo (7.87 in./mo)}$$

Based on nitrogen loading,

$$LA = \frac{Cp(PR - ET) + NU(10)}{(1 - f)Cn - Cp}$$

$$\therefore LA = \frac{10(12 - 7) \times 12 + 200(10)}{(1 - 0.2)40 - 10}$$

$$\therefore LA = 118.2 \text{ cm/a} = 0.0958 \text{ m/mo (3.88 in./mo)}$$

The land area requirement will be calculated using the lowest of the above two values (i.e., 9.85 cm/mo):

$$\therefore \text{Area for disposal of wastewater} = \frac{X \times 365/10\,000}{9.85/100 \times 12} = 123.52 \text{ ha}$$

$$\therefore X = 4000 \text{ m}^3/\text{d (733.9 gpm)}$$

Hence, 4000 m³/d (733.9 gpm) of wastewater can be treated with the designed land treatment system.

Chapter 24
PHYSICAL–CHEMICAL TREATMENT

Problems

PROBLEM 24.1

Explain the difference between flocculation and coagulation.

PROBLEM 24.2

A tertiary dual-media filter treats the secondary effluent from an advanced water resource recovery facility to remove suspended solids and phosphorus. Alum is added upstream of the filter to precipitate phosphorus and polymer is added as coagulant. The filters were designed based on a surface loading rate of 0.003 m³/m²·s (4.4 gpm/sq ft). Backwashes were designed to occur every 24 hours or when the head loss through the filter exceeded 0.6 m (2 ft). The backwash flowrate has been measured to be 13.5 L/s (20 gpm) and results in a filter bed expansion of between 25 and 30%. Operators at the facility have noted that the backwash frequency has increased and backwashes are occurring every 6 hours. Which of the following is least likely to be the cause of the increase in backwash frequency?

- **a)** The surface wash system has failed.
- **b)** The backwash rate is too low.
- **c)** The filter underdrains are plugged.
- **d)** The concentration of suspended solids in the feed to the filters has increased from the typical value of 10 to 35 mg/L.
- **e)** The dosage of alum applied for phosphorus removal is too high.

PROBLEM 24.3

In tapered flocculation, the mixing intensity is decreased in each zone of the flocculation tank as the flow proceeds through the tank from the entrance to the exit. Explain why this process improves the removal of suspended solids in a settling tank.

PROBLEM 24.4

Raw wastewater entering the primary clarifiers of a conventional activated sludge facility contains 200 mg/L of total suspended solids (TSS). There are two rectangular primary clarifiers, each 10-m (32.8-ft) wide and 15-m (49.2-ft) long. At an average daily flow of 10 000 m³/d (2.64 mgd), the primary effluent from the clarifiers contains an average of 100 mg/L TSS. Operating staff want to improve the performance of the primary clarifiers and have started to add an organic polymer to the primary clarifier influent at a dosage of 1 mg/L. Monitoring of the performance of the clarifiers has shown that the clarifier performance has improved and, on average, 70% removal of the raw wastewater TSS is being achieved. The sludge pumped from the primary clarifiers has a suspended solids concentration of 3% and did not increase after polymer addition was started. What volume of sludge must the operators pump from the clarifiers every day to prevent the sludge from accumulating?

    a)   The operators should pump the same amount of sludge as was pumped before polymer addition was started.

    b)   The operators should pump twice as much sludge as was pumped before polymer addition was started.

    c)   The operators should pump 20.0 m³/d (5284 gpd) of sludge.

    d)   The operators should pump 66.7 m³/d (17 622 gpd) of sludge.

    e)   The operators should pump 1.4 times as much sludge as was pumped before polymer addition was started.

PROBLEM 24.5

A tertiary filter treating secondary effluent from a conventional activated sludge facility reduces the suspended solids concentration in the secondary effluent from 10 to 1 mg/L. The total feed into the filters averages 5000 m³/d (1.32 mgd). The total backwash flow from the filters is 5% of the filter feed flow. What is the concentration of suspended solids in the backwash flow?

PROBLEM 24.6

The effluent from an advanced water resource recovery facility is treated in granular activated carbon (GAC) adsorption columns to remove refractory organics that are not biodegraded in the activated sludge process. The secondary effluent contains 10 mg/L of total organic carbon (TOC) and the GAC columns remove 90% of the TOC. The average flow to the GAC columns is 10 000 m³/d (2.64 mgd). The total GAC in the contactors is 300 kg (661.5 lb). Laboratory adsorption tests have determined that the GAC has a capacity to adsorb 5 kg of TOC per kg of GAC (5 lb of TOC per lb of GAC). How many days can the GAC contactors be operated before the GAC becomes saturated and the GAC must be regenerated or replaced?

PROBLEM 24.7

An extended aeration activated sludge water resource recovery facility is required to remove phosphorus from its discharge to a concentration of 1 mg/L. Chemical precipitation of the phosphorus is considered to be the easiest process to retrofit to the existing facility. Measurements made at the facility have shown that, of the 8 mg/L of total phosphorus entering the facility in the raw wastewater, approximately 4 mg/L is associated with particulate matter and the remainder is soluble. Chemicals will be added to the flow entering the secondary clarifiers to precipitate phosphorus and the precipitated phosphorus will be settled with the mixed liquor in the clarifiers and wasted to the facility digesters with the waste activated sludge. Operating staff are considering adding several chemicals, including alum, ferric chloride, polymer, sodium aluminate, and lime. Jar testing is being scheduled to determine the optimum chemical dosage and the relative cost of each chemical. Which chemicals should the operating staff not test during the jar tests?

PROBLEM 24.8

The raw wastewater flow to a conventional activated sludge water resource recovery facility varies between 5000 and 15 000 m³/d (1.32 and 3.96 mgd). Commercial liquid alum (49% alum) will be used for phosphorus removal. Commercial liquid alum (49% dry alum) contains 0.66 kg alum/L solution (5.4 lb dry alum/gal). Jar testing has shown that the alum dosage required to achieve a reduction in the total phosphorus concentration from 8 to 1 mg/L is between 50 and 75 mg/L as $Al_2(SO_4)_3 \cdot 14H_2O$. The alum feed pump speed will be automatically controlled so that the alum dosage is changed in proportion to the flowrate. Operators will manually adjust the dosage between 50 and 75 mg/L, depending on the phosphorus removal achieved. Operating staff are now preparing a specification for the alum feed pump. What range of alum pumping rate should be specified to the pump suppliers?

PROBLEM 24.9

Phosphorus can be removed from wastewater either chemically by the addition of metal salts such as alum or ferric chloride or biologically by appropriate design of the bioreactor. Which of the following is a true statement in comparing chemical phosphorus removal with biological phosphorus removal?

a) A chemical phosphorus removal process produces less sludge than a biological phosphorus removal process.

b) A chemical phosphorus removal process produces a sludge containing lower heavy metal content than a biological phosphorus removal process.

c) It is more difficult to retrofit a chemical phosphorus removal process to an existing activated sludge facility than a biological phosphorus removal process.

d) Chemical phosphorus removal is a simpler process to control than a biological phosphorus removal process.

e) Chemical phosphorus removal depends on the activity of a particular bacterium to be effective.

PROBLEM 24.10

Refractory organics are

- **a)** Rapidly biodegradable organic matter.
- **b)** Measured by total biochemical oxygen demand (BOD).
- **c)** Measured by carbonaceous BOD.
- **d)** Organic matter not removed by the treatment process.
- **e)** All of the above

PROBLEM 24.11

What concerns arise around solids handling when coagulation or flocculation chemicals are used?

PROBLEM 24.12

What is the purpose of ballasted media?

PROBLEM 24.13

To what pH range must water be reduced to after lime addition to permit the precipitation of calcium carbonate?

- **a)** 10 to 9.5
- **b)** <8.8
- **c)** <7

PROBLEM 24.14

What are two important process control parameters for coagulant dosing that can be easily measured in-line?

PROBLEM 24.15

What are the two main process measurements that indicate that it is time to initiate a filter backwash?

PROBLEM 24.16

Which of the following is not a physical–chemical process for ammonia reduction?

a) Ammonia stripping

b) Ion exchange

c) Anammox

Solutions

SOLUTION 24.1

Coagulation is the process of stabilizing the surface charge on particulate matter by the addition of an inorganic or an organic chemical (a coagulant). This neutralizes the mutual charge on the particles, which previously hindered settling. Flocculation is the process of agglomerating the destabilized particulate matter by the addition of a flocculent chemical to bind the particles into "chains" called *floc*. This will enhance its removal by physical processes such as sedimentation or filtration. If both processes are used, the coagulation occurs before flocculation in a treatment process train.

SOLUTION 24.2

b)   The backwash rate is too low.

If the backwash rate is sufficient to achieve a 25 to 30% expansion of the media in the filter bed, it is likely that sufficient scouring is occurring and the backwash rate is appropriate for the system and not the cause of the increased backwash frequency.

SOLUTION 24.3

In the flocculation process, a lower mixing intensity (i.e., a lower velocity gradient) produces a larger floc. In the tapered flocculation process, as the mixing intensity decreases through the tank, the particle size increases. The settling rate of a particle increases as the particle size increases. Therefore, the larger particles produced by the tapered flocculation process will settle more quickly in the settling tank, improving the removal efficiency of the settling process. Tapered flocculation also reduces the potential for floc breakup.

SOLUTION 24.4

e) The operators should pump 1.4 times as much sludge as was pumped before polymer addition was started.

The amount of solids removed after the addition of polymer is $(0.70 \times 200$ mg/L$) =$ 140 mg/L. Before polymer addition, the amount of solids removed was $(200$ mg/L $-$ 100 mg/L$) = 100$ mg/L. Therefore, the amount of solids removed has increased by $(140/100) = 1.4$ times. Because the concentration of the sludge produced by the primary clarifiers has not changed, the volume of sludge pumped will need to increase by the same amount as the quantity of solids removed has increased.

SOLUTION 24.5

180 mg/L TSS

In International Standard units:

The amount of suspended solids removed by the filter from a flow of 5000 m³/d is $(10$ mg/L $- 1$ mg/L$) = 9$ mg/L. These solids are concentrated in a backwash flow of 5% of 5000 m³/d $= 250$ m³/d. Therefore, the concentration of suspended solids in the backwash is as follows:

$$(9 \text{ mg/L} \times 5000 \text{ m}^3/\text{d})/250 \text{ m}^3/\text{d} = 180 \text{ mg/L}$$

In U.S. customary units:

The backwash volume is 5% of 1.32 mgd $= 0.066$ mgd

$$(9 \text{ mg/L} \times 1.32 \text{ mgd})/(0.066 \text{ mgd}) = 180 \text{ mg/L}$$

SOLUTION 24.6

16.7 days

$$\text{GAC columns remove } (0.90 \times 10 \text{ mg/L}) \text{ of TOC} = 9 \text{ mg/L}$$

In International Standard units:

$$\text{The mass of TOC removed each day} = 9 \text{ mg/L} \times 10\,000 \text{ m}^3/\text{d} \times$$
$$1000 \text{ L/m}^3 \times 1 \text{ g}/1000 \text{ mg} \times 1 \text{ kg}/1000 \text{ g} = 90 \text{ kg TOC/d}$$

The TOC adsorption capacity of the GAC = 300 kg GAC × 5 kg TOC/kg GAC

= 1500 kg TOC

Therefore, the GAC columns will require regeneration or replacement as follows:

1500 kg TOC/90 kg TOC/d = 16.7 days

In U.S. customary units:

Capacity of GAC = 661.5 lb × 5 lb TOC/lb GAC = 3307.5 lb

Regeneration needed in 3307.5 lb/198 lb/d = 16.7 days

SOLUTION 24.7

Operating staff should not test lime and polymer. Lime is only effective for phosphorus precipitation at a high pH. If added to the mixed liquor, the high pH will adversely affect the activity of the biomass. Polymer will coagulate the particulate matter containing phosphorus, but will not precipitate any soluble phosphorus. Because soluble phosphorus represents a significant fraction of the total phosphorus in the wastewater, polymer alone cannot achieve the target effluent total phosphorus concentration of 1 mg/L. Alum, ferric chloride, or sodium aluminate can be used and the jar tests will determine which is the most cost-effective.

SOLUTION 24.8

Approximately 250 to 1200 mL/min (0.05 to 0.40 gpm)

In International Standard units:

The minimal amount needed will be at the low dose and low flowrate:

$$50\frac{mg}{L} \times \frac{1\,L}{0.66\,kg} \times \frac{5000\,m^3}{d} \times \frac{1\,kg}{1\,000\,000\,mg} \times \frac{1000\,L}{1\,m^3} = 378\ L/d$$

The maximum amount needed will be at the high dose and the high flowrates:

$$\frac{75\,mg/L \times (1\,L)}{(0.66\,kg)} \times \frac{(15\,000\,m^3)}{d} \times \frac{1\,kg}{1\,000\,000\,mg} \times \frac{1000\,L}{1\,m^3} = 1705\ L/d$$

379 to 1704 L/d × 1000 mL/L × d/1440 min = 263 to 1183 mL/min

Therefore, an alum feed pump covering a range of approximately 250 to 1200 mL/min should be specified.

In U.S. customary units:

$$102 \text{ gpd to } 459 \times d/1440 \text{ min} = 0.070 \text{ gpm to } 0.32$$

Therefore, an alum feed pump covering a range of approximately 0.05 to 0.40 gpm should be specified.

SOLUTION 24.9

d) Chemical phosphorus removal is a simpler process to control than a biological phosphorus removal process.

SOLUTION 24.10

d) Organic matter not removed by the treatment process.

SOLUTION 24.11

The addition of chemicals for coagulation typically increase sludge volumes. They may also precipitate heavy metals, affecting the biosolids management options dependent on local heavy metal concentration limitations. Certain coagulants may also affect the dewatering efficiency. The thoughtful selection of a coagulant that had manageable side effects can minimize the effect on the facility's operation.

SOLUTION 24.12

Ballasted media is used in high-rate clarification. It is the media such as microsand to which the floc is fixed to with the addition of coagulant and a polymer. The microsand settles very quickly, allowing a much smaller tank size to be used. The settled media is then recycled and used again.

SOLUTION 24.13

a) 10 to 9.5

SOLUTION 24.14

Temperature and pH.

SOLUTION 24.15

Headloss through the filter and turbidity of the effluent. If there is too much headloss with no associated turbidity increase, it may indicate that the filter is clogged and not operating efficiently and could indicate too high a polymer dosing. Conversely, a filter that has turbidity breakthrough without headloss may have too low a polymer dose.

SOLUTION 24.16

c)   Anammox

Anammox is a biological method for ammonia reduction.

Chapter 25
PROCESS PERFORMANCE IMPROVEMENTS

Problems

PROBLEM 25.1

Why is it important to review the water resource recovery facility's (WRRF's) historical operational data when establishing process performance targets?

a) To determine historical unit process loading condition and performance
b) To identify factors that may limit facility performance
c) To provide information for planning other process performance improvement tasks
d) All of the above

PROBLEM 25.2

Hydraulic capacity is determined by which of the following?

a) Structure elevation
b) Pump capacity
c) Pipe and fitting size
d) Configuration
e) All of the above

PROBLEM 25.3

Some of the most common sources of error in a solids mass balance are the following:

a) Representative sample collection
b) Accurate flow measurement
c) Effect of recycle streams
d) All of the above

PROBLEM 25.4

Manipulation of the hydraulic model can be used to predict performance.

a) True
b) False

PROBLEM 25.5

To accurately calibrate the hydraulic model, verification of the key hydraulic elements is necessary.

a) True
b) False

PROBLEM 25.6

Process modifications to decrease the oxygen demand to the aeration system and increase capacity could include which of the following?

a) Pre-precipitation in the primary clarifiers
b) Increasing the food-to-microorganism ratio
c) Increasing the solids recycle streams
d) All of the above

PROBLEM 25.7

Flow equalization of primary effluent can be used to reduce both the oxygen demand and energy requirements of the secondary system.

a) True
b) False

PROBLEM 25.8

Installing anoxic selector zones to the design of an existing nitrifying aeration tank can, in effect, decrease the demands placed on the blower.

a) True
b) False

PROBLEM 25.9

Flow metering equipment commonly found in WWRFs can be divided into two basic categories: open-channel flow meters and full-pipe flow meters.

a) True
b) False

PROBLEM 25.10

For diffusers, the oxygen-transfer efficiency is affected by both temperature and pressure.

a) True
b) False

PROBLEM 25.11

Offgas analysis is used to calculate the oxygen transferred to the mixed liquor.

a) True
b) False

PROBLEM 25.12

The simplest and most reliable method of checking flow meter accuracy is a fill-and-draw test.

a) True
b) False

Solutions

SOLUTION 25.1

d) All of the above

Historical data review is an essential component of a process performance improvement project. It provides the opportunity to become familiar with the operation and performance of the WRRF. The general objectives of the historical data review are to

- Determine historical unit process loading condition and performance,
- Identify factors that may limit facility performance, and
- Provide information for planning other process performance improvement tasks.

SOLUTION 25.2

e) All of the above

Hydraulic capacity is determined by structure elevations, hydraulic control section elevations, channel arrangements, pump capacity, and pipe and fitting sizes.

SOLUTION 25.3

c) Effect of recycle streams

Some of the most common sources of error in a solids mass balance are

- Nonrepresentative samples (analytical accuracy and sampling techniques);
- Inaccurate flow monitoring;
- Effect of periodic recycle streams (the boundaries of the balance must be clearly defined and all inputs/outputs of the defined boundaries must be accounted for in the mass balance);
- Assumptions made concerning accumulations; and
- Time differences between sampling.

SOLUTION 25.4

a) True

Hydraulic modeling is an important component of a process performance evaluation and improvement project. Hydraulic modeling entails three main steps:

(1) Development of the mathematical model,

(2) Calibration, and

(3) Manipulation of the model to predict performance.

SOLUTION 25.5

a) True

Once the model is constructed, field verification and calibration are required. Field verification consists of measuring the flow and water surface level (hydraulic grade line) at key locations in the facility.

SOLUTION 25.6

a) Pre-precipitation in the primary clarifiers

The following process modifications can potentially reduce oxygen demand to the aeration system:

- Pre-precipitation in the primary clarifiers,
- Retrofitting the primary clarifiers,
- Flow equalization,
- Solids recycle stream control, and
- Denitrification in the aeration basins.

SOLUTION 25.7

a) True

Flow equalization of primary effluent can be used to reduce both the oxygen demand and energy requirements of the secondary system.

SOLUTION 25.8

a) True

When the existing aeration basins have enough volume to allow for installation of anoxic zones for denitrification, the oxygen bound in nitrates and nitrites can be used for further oxidation of organic matter. In this way, savings in oxygen requirements (and related aeration energy) of up to 10 to 20% can be achieved.

SOLUTION 25.9

a) True

SOLUTION 25.10

a) True

SOLUTION 25.11

a) True

Offgas analysis measures the composition of the air entering an aeration tank and the off-gas exiting an aeration tank. By comparing the oxygen content of these two gas streams, it is possible to calculate the overall oxygen transferred to the mixed liquor.

SOLUTION 25.12

a) True

The simplest and most reliable method of checking flow meter accuracy is a fill-and-draw test. The test involves drawing down the liquid level in a basin or tank and filling it back up while recording the meter reading.

Chapter 26
EFFLUENT DISINFECTION

Problems

PROBLEM 26.1

Which of the following best describes Chick's Law in words?

The rate of change of the number of organisms is

a) directly proportional to the number of organisms.
b) inversely proportional to the number of organisms.
c) constant.
d) random.
e) None of the above

PROBLEM 26.2

The two basic parameters that affect disinfection efficiency are which of the following?

a) Mixing and dispersion
b) Cell resistance and clumping
c) Time of exposure and concentration of disinfectant
d) Oxidation and complexation
e) None of the above

PROBLEM 26.3

Ultraviolet disinfection works not by directly killing exposed microorganisms, but by destroying their ability to reproduce.

a) True
b) False

PROBLEM 26.4

As the operator of a large water reclamation facility using UV disinfection, you have recently noticed several effluent fecal coliform measurements exceeding the design effluent limit of 200 CFU/100 mL. Which of the following is the most likely cause of the problem?

a) The effluent transmittance is higher than the design value

b) The level in one of the UV channels is higher than the design level

c) Your system does not have automatic lamp wipers and you have not cleaned the lamps in the last 2 months

d) a) and/or b), but not c)

e) b) and/or c), but not a)

PROBLEM 26.5

A low-pressure, low-intensity UV system is designed for a secondary treated effluent with a peak flow of 94.75 ML/d (25 mgd). The UV system consists of two channels in parallel (assume equal flow split between two channels), each equipped with three banks of UV lamps. Each bank contains 16 modules arranged side by side, with each module containing 16 lamps stacked vertically. One bank in each channel is a redundant standby. The lamps are oriented parallel to the direction of flow and are arranged in a square grid in the cross-section, with a center-to-center distance of 76.2 mm (3 in.) in both the vertical and horizontal directions. The overall cross-section in each channel, including the lamp grid and effluent flow, measures 1219.2-mm (48-in.) across by 1219.2-mm (48-in.) deep. Each lamp has a sleeve outside diameter of 38.1 mm (1.5 in.) and an effective arc length of 1828.8 mm (72 in.).

The firm UV contact time for this system at peak flow is which of the following?

a) 20 minutes

b) 8 seconds

c) 12 seconds

d) 2 seconds

PROBLEM 26.6

Using the information from Problem 26.5, if the effluent total suspended solids (TSS) is 20 mg/L, the UV transmittance is 60%, and a UV dose of 30 mW-s/cm² is needed to provide the desired level of inactivation, what average intensity will the UV system need to deliver?

a) 1.30 mW/cm^2

b) 2.60 mW/cm^2

c) 3.75 mW/cm^2

d) 7.98 mW/cm^2

PROBLEM 26.7

The primary mechanism by which chlorine disinfection destroys essential cellular material and processes is which of the following?

a) Hydrolysis
b) Acid formation
c) Dissociation
d) Complexation
e) None of the above

PROBLEM 26.8

There is no limit to the withdrawal rate of gaseous chlorine from pressurized cylinders and containers without the use of evaporators.

a) True
b) False

PROBLEM 26.9

A wastewater treatment utility is required to maintain a free chlorine residual of 5 mg/L at the end of the chlorine contact tanks. The initial chlorine demand (not including ammonia) is estimated to be 2 mg/L, and it may be assumed that each milligram of ammonia-nitrogen exerts a chlorine demand of 10 mg/L. The effluent to be disinfected has an ammonia-nitrogen concentration of 0.5 mg/L. All chlorine quantities (doses and residuals) are expressed on an "as Cl" basis.

What is the breakpoint chlorine dose?

a) 2 mg Cl/L
b) 5 mg Cl/L
c) 7 mg Cl/L
d) 12 mg Cl/L
e) Other

PROBLEM 26.10

Using the information from Problem 26.9, what is the total chlorine dose required?

a) 2 mg Cl/L
b) 5 mg Cl/L
c) 7 mg Cl/L
d) 12 mg Cl/L
e) Other

PROBLEM 26.11

Using the information from Problem 26.9, if the chloramine residual is 2 mg Cl/L, what is the total chlorine residual?

a) 2 mg Cl/L
b) 5 mg Cl/L
c) 7 mg Cl/L
d) 12 mg Cl/L
e) Other

PROBLEM 26.12

Using the information from Problem 26.9, if the peak effluent flowrate is 37.9 ML/d (10 mgd), and assuming 100% gaseous feed chlorine purity, what is the peak chlorine gas mass feed rate?

a) 455 kg/d (1001 lb/d)
b) 265 kg/d (584 lb/d)
c) 910 kg/d (2002 lb/d)
d) 190 kg/d (417 lb/d)
e) Other

PROBLEM 26.13

Using the information from Problem 26.9, if the pH of the chlorinated effluent at the end of the chlorine contact tanks is 6.0, the free chlorine residual exists as which of the following:

a) Predominantly HOCl
b) Predominantly as OCl_2
c) Significant concentrations of both HOCl and OCl_2
d) Reverts back to Cl_2 gas
e) None of the above

PROBLEM 26.14

How many milligrams of sodium bisulfite are theoretically required to dechlorinate a free chlorine residual of 1 mg as Cl? Support your answer with a balanced chemical equation.

a) 1.0
b) 2.9
c) 3.5
d) 5.0
e) Other

PROBLEM 26.15

Disinfection by ozone occurs through which of the following mechanisms?

a) Direct oxidation/destruction of the cell wall with leakage of cellular constituents outside of the cell
b) Reactions with radical byproducts of ozone decomposition
c) Damage to the constituents of the nucleic acids (purines and pyrimidines) inside the cell, which alters their cellular genetic material, thereby preventing cell replication
d) All of the above
e) None of the above

Solutions

SOLUTION 26.1

a)   directly proportional to the number of organisms.

This is essentially the verbal equivalent of Chick's Law, which is expressed as follows:

$$dN/dt = -kN$$

Where
dN/dt = rate of change of the number of organisms per unit volume,
k = organism inactivation rate constant, and
N = number of surviving organisms per unit volume.

SOLUTION 26.2

c)   Time of exposure and concentration of disinfectant

These are the classic components of disinfection efficiency. Increasing one or both increases the level of inactivation achieved.

SOLUTION 26.3

a)   True

Ultraviolet radiation at a wavelength of 253.7 nm alters the cellular genetic material of microorganisms in such a way that it cannot be replicated. The microorganisms cannot reproduce and are thus inactivated.

SOLUTION 26.4

e)   b) and/or c), but not a)

Higher effluent transmittance a) would actually result in an increase in the UV intensity delivered to the microorganisms and, therefore, reduce the effluent fecal coliform count. In contrast, a higher level in the UV channel b) would mean that the portion of the effluent flowing through the top layers is farther away from the UV source and is receiving lower UV intensity. Similarly, lamp sleeve fouling or coating resulting from not cleaning the lamps for an extended period c) would reduce the delivered UV intensity. Either b) or c) would likely result in higher effluent fecal coliform counts.

SOLUTION 26.5

b)   8 seconds

In International Standard units:

Consider the square cross-section defined by the centerlines of any four adjacent lamps:

$$s_{lamp} \text{ (Lamp spacing)} = \text{Length of each side of square} = 76.2 \text{ mm}$$

$$d_{lamp} = \text{Lamp sleeve outer diameter} = 38.1 \text{ mm}$$

$$a_{total} = \text{Total area of defined cross-section} = S_{lamp}{}^2 = 5806 \text{ mm}^2$$

$$a_{lamp} = \text{Area occupied by lamp} = [\pi(d_{lamp})^2]/4 = [3.14(38 \text{ mm})^2]/4 = 1140 \text{ mm}^2$$

$$a_{flow} = \text{Flow cross-sectional area per lamp} = a_{total} - a_{lamp} = 5806 \text{ mm}^2 - 1140 \text{ mm}^2$$
$$= 4666 \text{ mm}^2$$

$$L = \text{Lamp arc length} = 1828.8 \text{ mm}$$

$$v_{flow} = \text{Contact volume per lamp} = a_{flow}L = 8.53 \text{ L}$$

$$n_{duty} = \text{Number of duty lamps} = 2 \text{ channels} \times 2 \text{ duty banks per channel} \times 16 \text{ modules}$$
$$\text{per bank} \times 16 \text{ lamps per module} = 1024$$

$$V_{flow} = \text{Total firm contact volume} = v_{flow} \times n_{duty} = 8735 \text{ L}$$

$$Q_{peak} = \text{Peak effluent flow} = 94.75 \text{ ML/d}$$

$$T = \text{Total firm contact time} = V_{flow}/Q_{peak} = [(8735 \text{ L})/(94\,750\,000 \text{ L/d})] \times (86\,400 \text{ s/d}) =$$
$$8 \text{ seconds}$$

In U.S. customary units:

Consider the square cross section defined by the centerlines of any four adjacent lamps:

$$s_{lamp} \text{ (Lamp spacing)} = \text{Length of each side of square} = 3 \text{ in.}$$

$$d_{lamp} = \text{Lamp sleeve outer diameter} = 1.5 \text{ in.}$$

$$a_{total} = \text{Total area of defined cross-section} = S_{lamp}{}^2 = 9 \text{ sq in.}$$

$$a_{lamp} = \text{Area occupied by lamp} = [\pi(d_{lamp})^2]/4 = [3.14(0.75 \text{ in.})^2]/4 = 1.766 \text{ sq in.}$$

$$a_{flow} = \text{Flow cross-sectional area per lamp} = a_{total} - a_{lamp}$$
$$= 9 \text{ sq in.} - 1.766 \text{ sq in.} = 7.234 \text{ sq in.}$$

$$L = \text{Lamp arc length} = 72 \text{ in.}$$

$$v_{flow} = \text{Contact volume per lamp} = a_{flow}L = 520.8 \text{ cu in.}$$

$$n_{duty} = \text{Number of duty lamps} = 2 \text{ channels} \times 2 \text{ duty banks per channel}$$
$$\times 16 \text{ modules per bank} \times 16 \text{ lamps per module} = 1024$$

$$V_{flow} = \text{Total firm contact volume} = v_{flow} \times n_{duty} = 533\ 299 \text{ cu in.} = 2309 \text{ gal}$$

$$Q_{peak} = \text{Peak effluent flow} = 25 \text{ mgd}$$

$$T = \text{Total firm contact time} = V_{flow}/Q_{peak} = \frac{(2309 \text{ gal})}{(25\ 000\ 000 \text{ gpd})} \times 86\ 400 \text{ s/d} = 8 \text{ seconds}$$

SOLUTION 26.6

c) 3.75 mW/cm^2

$$\text{UV dose (D)} = \text{Intensity (I)} \times \text{Contact time (T)}$$

$$\text{Therefore, I} = D/T = (30 \text{ mW-s/cm}^2)/8 \text{ s} = 3.75 \text{ mW/cm}^2$$

Note: The effluent TSS and transmittance information provided is extraneous for the purposes of this solution.

SOLUTION 26.7

e) None of the above

The main disinfection mechanism by which chlorine destroys cellular material is oxidation. Chlorine is a strong oxidant and disrupts cellular processes by oxidizing essential proteins and enzymes. Although some of the other processes mentioned in the question do occur as part of aqueous chlorine chemistry, they are not the primary disinfection mechanism. The primary chlorine disinfection reactions are oxidation–reduction reactions.

SOLUTION 26.8

b) False

Unless supplemental heat is provided via devices such as evaporators, the rate of withdrawal of gaseous chlorine from pressurized cylinders and containers is limited by the rate at which heat can be transferred to the container from the surrounding atmosphere for the purpose of vaporizing the liquid chlorine to gas.

SOLUTION 26.9

c) 7 mg Cl/L

Breakpoint chlorine demand is the sum of the initial chlorine demand (not including ammonia-related demand) plus the chloramination demand exerted by ammonia.

$$\text{Chloramination demand} = 0.5 \text{ mg/L ammonia-N} \times 10 \text{ mg}$$
$$\text{Cl per mg ammonia-N} = 5 \text{ mg Cl/L}$$

$$\text{Breakpoint chlorine demand} = 2 \text{ mg Cl/L (initial)} + 5 \text{ mg}$$
$$\text{Cl/L (ammonia)} = 7 \text{ mg Cl/L}$$

SOLUTION 26.10

d) 12 mg Cl/L

$$\text{Total required dose} = \text{Breakpoint dose} + \text{Free residual required} =$$
$$7 \text{ mg Cl/L} + 5 \text{ mg Cl/L} = 12 \text{ mg Cl/L}$$

SOLUTION 26.11

c) 7 mg Cl/L

$$\text{Total residual} = \text{Combined (chloramine) residual} + \text{Free residual}$$
$$= 2 \text{ mg Cl/L} + 5 \text{ mg Cl/L} = 7 \text{ mg Cl/L}$$

SOLUTION 26.12

c) 910 kg/d (2002 lb/d)

The key step here is conversion from "as Cl" basis to "as Cl_2" basis. All chlorine doses and residuals are expressed "as Cl". However, gaseous chlorine is molecular chlorine (Cl_2). When dissolved in water, one of the atoms from the molecule is oxidized to form hypochlorite ion, while the other atom is reduced to form chloride ion. Only the hypochlorite provides disinfection. Thus, on a mass basis, to provide a chlorine dose of X mg/L as Cl, a gaseous mass flowrate of $2X$ mg/L as Cl_2 is required.

$$\text{Gaseous } Cl_2 \text{ feed rate required} = 12 \text{ mg/L (chlorine dose as Cl)} \times 2 = 24 \text{ mg/L}$$

In International Standard units:

$$\text{Gaseous Cl}_2 \text{ mass feed rate required} = 37.9 \text{ ML/d} \times 24 \text{ mg/L} = 910 \text{ kg/d}$$

In U.S. customary units:

$$\text{Gaseous CL2 mass feed rate required} = 10 \text{ mgd} \times 24 \text{ mg/L} \times \left(\frac{8.34 \text{ lb/mil. gal}}{\text{mg/L}} \right) = 2002 \text{ lb/d}$$

SOLUTION 26.13

    a)   Predominantly HOCl

This is based on the equilibrium of reversible acid–base reactions as a function of pH.

SOLUTION 26.14

    b)   2.9

The balanced chemical equation for dechlorination of free chlorine residual by sodium bisulfite is as follows:

$$\text{NaHSO}_3 + \text{HOCl} \leftrightarrow \text{NaHSO}_4 + \text{HCl}$$

$$(23 \text{ mg} + 1 \text{ mg} + 32 \text{ mg}) + (16 \text{ mg} \times 3 \text{ mg}) =$$
$$104 \text{ mg of sodium bisulfite to reduce } 35.5 \text{ mg of free chlorine residual as Cl}$$

The required ratio, therefore, $= 104 \text{ mg}/35.5 \text{ mg} = 2.9$

SOLUTION 26.15

    d)   All of the above

Chapter 27
MANAGEMENT OF SOLIDS

Problems

PROBLEM 27.1

A water resource recovery facility's (WRRF's) biosolids program that meets best management practices (BMP)

a) requires a sampling and analysis plan that meets regulatory requirements, but does not need chain-of-custody forms, especially for facility samples.

b) requires a written sampling and analysis plan and has a detailed quality assurance/quality control (QA/QC) procedure for in-house as well as contract laboratory analysis.

c) requires sampling and analysis only when regulatory agencies require it or when there are complaints.

PROBLEM 27.2

What role does an industrial pretreatment program play in ensuring the quality of biosolids for beneficial reuse?

a) Ensures that the application of biosolids does not cause contamination of land or groundwater because of toxics or heavy metals

b) Ensures that industries cannot be held responsible for any contamination caused by their discharges

c) Ensures that the public cannot hold the WRRF management responsible for any contamination

PROBLEM 27.3

CFR40503 defined requirements for beneficial use of biosolids. The classes and specific requirement(s) are as follows:

a) Class A, B, and C. Pathogen reduction
b) Class A-EQ and B-EQ. Pathogen and volatile solids reduction
c) Class A and B. Pathogen reduction and volatile solids reduction

PROBLEM 27.4

When using land application as a disposal method it is not necessary to have another disposal alternative because land is always available.

a) True
b) False

PROBLEM 27.5

What is the difference between biosolids and sludge?

a) Sludge is the residual product from primary and secondary treatment before stabilization.
b) Biosolids are residuals from all processes within a WRRF.
c) Sludge and biosolids mean the same thing.

PROBLEM 27.6

What effect does the addition of lime for lime stabilization or bulking agents for composting have on the overall cost of biosolids treatment?

a) Reduces disposal costs because the solids are stabilized
b) Increases the cost of disposal because additional mass is added
c) Does not change the overall cost, but improves the product

PROBLEM 27.7

Calculate the mass of 100 kg (220 lb) of biosolids at each of the following solids concentrations: 5%, 24%, and 90%.

a) 2000 kg (4400 lb), 417 kg (917 lb), 111 kg (244 lb)
b) 5 kg (11 lb), 24 kg (22 lb), 90 kg (198 lb)
c) 50 kg (110 lb), 2400 kg (5289 lb), 9000 kg (19 800 lb)

PROBLEM 27.8

What is the best approach for public acceptance of land application of biosolids?

a) Explaining that biosolids do not have odors because they are stabilized
b) Producing public education materials and producing biosolids as odor free as possible
c) Not allowing citizen groups or media to land application sites because they might misinform the public

PROBLEM 27.9

Why is recordkeeping so important for a beneficial use program?

a) To comply with regulatory requirements and to document quality and characteristics for regulators and inspectors
b) Recordkeeping is not required for beneficial use, only for landfilling purposes
c) To use in case of litigation and consent orders

PROBLEM 27.10

How does the liability and responsibility of a municipality change when a private contractor is used for the transportation and land application of biosolids?

a) Municipality is exempt from all regulatory requirements
b) Municipality is exempt from all litigation
c) Municipality is still responsible for regulatory compliance and for the conduct of the contractor

PROBLEM 27.11

The National Biosolids Partnership encourages the use of an Environmental Management System (EMS). What is the value of an EMS?

    a) Improve production and use of biosolids

    b) Promote the use of incineration to reduce volume and mass of biosolids

    c) Manage the systems used for environmental compliance

PROBLEM 27.12

When calculating the total cost of a biosolids treatment and beneficial use program, the following must be included:

    a) Primary and secondary settling, thickening, and trucking costs

    b) Thickening, stabilization, dewatering, and transport to application sites

    c) Thickening, dewatering, and payments to outside contractor

Solutions

SOLUTION 27.1

    b) requires a written sampling and analysis plan and has a detailed quality assurance/ quality control (QA/QC) procedure for in-house as well as contract laboratory analysis.

SOLUTION 27.2

    a) Ensures that the application of biosolids does not cause contamination of land or ground water due to toxics or heavy metals

SOLUTION 27.3

    c) Class A and B. Pathogen reduction and volatile solids reduction

SOLUTION 27.4

    b) False

SOLUTION 27.5

a)   Sludge is the residual product from primary and secondary treatment before stabilization.

SOLUTION 27.6

b)   Increases the cost of disposal because additional mass is added

SOLUTION 27.7

a)   2000 kg (4400 lb), 417 kg (917 lb), 111 kg (244 lb)

SOLUTION 27.8

b)   Producing public education materials and producing biosolids as odor free as possible

SOLUTION 27.9

a)   To comply with regulatory requirements and to document quality and characteristics for regulators and inspectors

SOLUTION 27.10

c)   Municipality is still responsible for regulatory compliance and for the conduct of the contractor

SOLUTION 27.11

a)   Improve production and use of biosolids

SOLUTION 27.12

b)   Thickening, stabilization, dewatering, and transport to application sites

Chapter 28
CHARACTERIZATION AND SAMPLING OF SLUDGES AND RESIDUALS

Problems

PROBLEM 28.1

Of the various residuals, screening and grit are composed primarily of water, but contain dissolved and suspended solids that can range from 0.5 to 7% total solids in an unthickened state.

- **a)** True
- **b)** False

PROBLEM 28.2

Thickening, stabilization, and dewatering processes are used to make which of the following residuals acceptable for disposal?

- **a)** Return activated sludge and nitrate recycle
- **b)** Sludge and biosolids
- **c)** Sludge and screenings
- **d)** Biosolids and grit

PROBLEM 28.3

Which two of the following are typical thickening tests?

a) Settled sludge volume and sludge volume index
b) Sludge density index and settleable solids
c) Sludge volume index and total suspended solids
d) Solids capture rate and total dissolved solids

PROBLEM 28.4

Which of the following are the most typical current products for conditioning sludge?

a) Ferric and lime
b) Organic polymers
c) Inorganic polymers
d) Alum

PROBLEM 28.5

Pretreatment programs can reduce the amount of hazardous chemicals discharged to a treatment plant and contaminating sludge.

a) True
b) False

PROBLEM 28.6

Sampling in water resource recovery facilities is performed for a twofold purpose—
_____ and _____.

PROBLEM 28.7

To obtain a representative sample, the sample should be which of the following?

a) Grabbed from floating solids near the surface
b) Collected in a well-mixed area (avoiding areas that have a large amount of solids accumulation)
c) A composite of settled sludge from settleometers
d) A grab from the primary sludge pump sample port

PROBLEM 28.8

Which of the following best describes compliance samples?

a) Have a specific location, frequency of analysis, type of sample (grab, composite, etc.), parameters to be measured, and method of measurement

b) Are only required during swimming season

c) Designate the location, but the frequency and types of tests will be left to the plant manager

d) Designate the frequency of analysis, type of sample and test, but not the location

PROBLEM 28.9

Chain-of-custody forms are important in case of litigation.

a) True

b) False

PROBLEM 28.10

Which of the following references contains methods for sample preparation and preservation?

a) *Standard Methods for the Examination of Water and Wastewater*

b) *Standard Methods for the Examination of Sludge and Sediment*

c) *Standard Methods for the Testing of Water and Wastewater*

d) *Standard Methods for Water and Wastewater Treatment*

PROBLEM 28.11

"QA/QC" stands for which of the following?

a) Quality accuracy and quantity of control testing

b) Quality assurance and quality control

c) Quality accuracy and quality control

d) Qualification of analysts and qualification of control programs

PROBLEM 28.12

Recordkeeping is a vital function of a good sampling and testing program.

   a)   True

   b)   False

Solutions

SOLUTION 28.1

   b)   False

SOLUTION 28.2

   b)   Sludge and biosolids

SOLUTION 28.3

   a)   Settled sludge volume and sludge volume index

SOLUTION 28.4

   b)   Organic polymers

SOLUTION 28.5

   a)   True

SOLUTION 28.6

Sampling in water resource recovery facilities is performed for a twofold purpose—compliance and process control.

SOLUTION 28.7

   b)   Collected in a well-mixed area (avoiding areas that have a large amount of solids accumulation)

SOLUTION 28.8

a) Have a specific location, frequency of analysis, type of sample (grab, composite, etc.), parameters to be measured, and method of measurement

SOLUTION 28.9

a) True

SOLUTION 28.10

a) *Standard Methods for the Examination of Water and Wastewater*

SOLUTION 28.11

b) Quality assurance and quality control

SOLUTION 28.12

a) True

Chapter 29
THICKENING

Problems

PROBLEM 29.1

The purpose of sludge thickening is to decrease the volume of sludge and to reduce the cost of downstream processes.

    a) True
    b) False

PROBLEM 29.2

What are the three types of settling that occur in a gravity thickener?

    a) Flocculation, coagulation, and sedimentation
    b) Filtration, sedimentation, and centrifugation
    c) Discrete, hindered, and compaction

PROBLEM 29.3

Gravity belt, rotary drum, and centrifugal thickening can achieve normal solids concentrations of 6 to 8%, and solids concentrations can be as high as 12 to 14%.

    a) True
    b) False

PROBLEM 29.4

How are gravity thickener loading rates expressed?

a) By the type of sludge handled
b) By the solids loading rate and overflow rate
c) By primary sludge and waste activated sludge

PROBLEM 29.5

It is important to ensure that sludge underflow pumping velocities of at least 0.75 m/s (2.5 ft/sec) are maintained to prevent solids deposition in the sludge discharge lines.

a) True
b) False

PROBLEM 29.6

Calculate the hydraulic loading in terms of cubic meters per day per square meters (gallons/day per square foot) for a gravity thickener 7.6 m (25 ft) in diameter, with a surface area of approximately 45.5 m² (490 sq ft) and a flowrate of 817 m³/d (0.216 mgd).

a) 15 m³/(m²·d) (380 gpd/sq ft)
b) 18 m³/(m²·d) (440 gpd/sq ft)
c) 21 m³/(m²·d) (520 gpd/sq ft)

PROBLEM 29.7

Calculate the solids loading on a gravity thickener in terms of kilograms per day per square meters (pounds per day per square feet) that receives a sludge flowrate of 817 m³/d (0.216 mgd), with a total solids concentration of 7000 mg/L (0.7%).

a) 115 kg/m²·d (21 lb/d/sq ft)
b) 125 kg/m²·d (26 lb/d/sq ft)
c) 150 kg/m²·d (32 lb/d/sq ft)

PROBLEM 29.8

Typical parameters that are monitored for monitoring and controlling dissolved air floatation (DAF) units include the following:

a) Solids and hydraulic loading rates
b) Air-to-solids ratio
c) Air volume and pressure
d) All of the above

PROBLEM 29.9

The hydraulic loading on a DAF includes the feed rate plus recycle.

a) True
b) False

PROBLEM 29.10

The air-to-solids ratio (pound air/pound solids) affects the sludge rise rate and the float depth. The necessary air-to-solids ratio is typically between 0.02 and 0.04.

a) True
b) False

PROBLEM 29.11

Gravity belt thickeners work by filtering free water from conditioned solids by applying pressure caused by the belt tension setting to squeeze the filtrate through a porous belt.

a) True
b) False

PROBLEM 29.12

The thickening mechanism of a rotary drum thickener is similar to a gravity belt thickener because the separated solids are retained on a filter screen while the filtrate is allowed to drain away.

a) True
b) False

PROBLEM 29.13

Centrifuges use the principle of sedimentation to separate liquids from solids—the same principle as the clarifiers and thickeners in the wet end of the facility.

a) True
b) False

Solutions

SOLUTION 29.1

a) True

SOLUTION 29.2

c) Discrete, hindered, and compaction

SOLUTION 29.3

a) True

SOLUTION 29.4

b) Solids loading rates and overflow rate

SOLUTION 29.5

a) True

SOLUTION 29.6

b)   18 m³/m²·d (440 gpd/sq ft)

In International Standard units:

$$817 \text{ m}^3/\text{d} \div 45.5 \text{ m}^2 = 18 \text{ m}^3/\text{m}^2$$

In U.S. customary units:

$$0.216 \text{ mgd} \div 490 \text{ sq ft} = 440 \text{ gpd/sq ft}$$

SOLUTION 29.7

b)   125 kg/m²·d (26 lb/d/sq ft)

In International Standard units:

$$(817 \text{ m}^3/\text{d} \times 1000 \text{ L/m}^3 \times 7000 \text{ mg/L} \times \text{kg}/1\,000\,000 \text{ mg}) \div 45.5 \text{ m}^2 = 125 \text{ kg/m}^2 \cdot \text{d}$$

In U.S. customary units:

$$(0.216 \text{ mgd} \times 8.34 \text{ lb/gal} \times 7000 \text{ mg/L}) \div 490 \text{ sq ft} = 26 \text{ lb/d/sq ft}$$

SOLUTION 29.8

d)   All of the above

SOLUTION 29.9

a)   True

SOLUTION 29.10

a)   True

SOLUTION 29.11

b) False

Gravity belt thickeners work by filtering free water from conditioned solids by gravity drainage through a porous belt.

SOLUTION 29.12

a) True

SOLUTION 29.13

a) True

Chapter 30
ANAEROBIC DIGESTION

Problems

PROBLEM 30.1

Which of the following stages in anaerobic digestion controls the process?

a) Breakdown of solids complex organic compounds
b) Methane-forming bacteria that feed on organic acids
c) Extracellular enzymes that convert volatile fatty acids to methane
d) Conversion of organic fatty acids, alcohols, carbon dioxide, and ammonia to volatile acids

PROBLEM 30.2

Two criteria determine the anaerobic digester's capacity to process the sludge properly. One is the solids retention time. What is the other?

a) Temperature
b) Hydraulic detention time
c) Methane-forming bacteria
d) Organic loading

PROBLEM 30.3

A potential effect of adding ferrous chloride to the recirculated digester sludge upstream of the heat exchangers is which of the following?

a) Vivianite formation
b) Explosion potential
c) Plating of inorganic carbon
d) Struvite formation

PROBLEM 30.4

What class of biosolids must meet vector attraction reduction requirements?

a) Class A and B
b) Vector attraction reduction is a goal, not a requirement
c) Class B
d) Class A

PROBLEM 30.5

The volatile acid-to-alkalinity ratio should range from approximately 0.1 to 0.2. Corrective actions should be taken when the ratio exceeds which of the following?

a) Less than 0.2 mg/L
b) 1.0 to 1.5
c) 0.8 to 0.6
d) 0.3 to 0.4
e) >0.2

PROBLEM 30.6

The most common cause of digester foaming is organic overload, which results in greater quantities of volatile fatty acids being produced than can be converted to methane. The best course of action would be to reduce digester feed and which of the following?

a) Maintain digester mixing
b) Increase the digester temperature
c) Increase the gas flow to the waste gas burner
d) Add ferric chloride to force the excess organic loading into solution

PROBLEM 30.7

In completing morning rounds, an operator finds that the digester floating cover is slightly tilted to one side. No obvious obstructions are noticed between the wall and the cover rim skirt and the digester is not foaming. What other conditions may be causing the cover to list?

a) The sludge level is extending above the skirt.
b) Water has accumulated in the attic space.
c) A vacuum was created during sludge withdrawal.
d) There is excess pressure in the gas system.

PROBLEM 30.8

A water resource recovery facility has two identically sized primary digesters that are fed primary sludge and thickened waste activated sludge throughout the day. The digesters overflow to the secondary digester, which serves as a holding tank for the centrifuge dewatering units. Given the information below, calculate the primary digester digestion detention time in days.

Parameter	Value
Digester diameter	24 m (80 ft)
Digester side water depth	5.5 m (18 ft)
Digester cone	2 m (8 ft)
Primary sludge flow	194 m³/d (51 356 gpd)
Thickened waste activated sludge flow	244 m³/d (64 394 gpd)
Digested sludge total solids	2.81%
Digested sludge volatile solids	67.60%

PROBLEM 30.9

A 2-L sample of digested sludge is used to determine the pH dosage required for a sour digester. If the digester volume is 3400 m³ (900 000 gal) and 52 mg sodium hydroxide was used in the test, how much sodium hydroxide should be added to the digester?

PROBLEM 30.10

A water reclamation facility has two 24-m (80-ft) diameter anaerobic digesters. Given the following information, calculate the volatile solids reduction.

Parameter	Value
Thickened sludge percent solids	4.40%
Thickened sludge percent volatile solids	77%
Thickened sludge flow	1321 m³/d (349 000 gpd)
Digested sludge percent solids	2.81%
Digested sludge percent volatile solids	68%
Digested sludge transferred to holding tank	827 m³/d (218 400 gpd)

Solutions

SOLUTION 30.1

b)   Methane-forming bacteria that feed on organic acids

In most cases, methane-forming bacteria control the process. Methane formers are sensitive to environmental factors (high ammonia concentrations, low phosphorus concentrations, low pH, temperature, or the presence of toxic substances) and reproduce slowly. Consequently, methane formers are difficult to grow and are easily inhibited. Process design and operation of conventional anaerobic digestion are, therefore, tailored to satisfy the needs of methane-forming bacteria.

SOLUTION 30.2

d)   Organic loading

SOLUTION 30.3

a)   Vivianite formation

When iron is added directly upstream of heat exchangers, vivianite can accumulate on the surfaces of the heat exchangers, causing plugging and decreasing the rate of heat transfer.

SOLUTION 30.4

a)   Class A and B

SOLUTION 30.5

d)   0.3 to 0.4

SOLUTION 30.6

a)   Maintain digester mixing

Organic overload can be minimized by feeding the digesters continuously (or as often as possible), blending different feed sludges well before feeding, ensuring that the digester mixing system is operable, and limiting the quantity of grease or scum in the digester feed.

SOLUTION 30.7

b)   Water has accumulated in the attic space.

The operator should check floating covers periodically to ensure that they are level and move freely. A tilted cover often reflects an uneven loading condition resulting from water accumulation in the attic space, uneven snow buildup, or a binding condition between the wall and the cover rim skirt. Excessive foam can also be identified during the cover inspection.

SOLUTION 30.8

12.7 days

$$\text{Digester detention time} = \text{Digester volume}/\text{Digester feed sludges}$$

In International Standard units:

$$\text{Volume} = \pi(\text{radius, m})^2(\text{height, m}) + \tfrac{1}{3}\pi(\text{radius, m})^2(\text{height, m})$$

$$(3.14)(12\text{ m})^2(5.5\text{ m}) + \tfrac{1}{3}(3.14)(12\text{ m})^2(2\text{ m})$$

$$2487\text{ m}^3 + 301\text{ m}^3 = 2788\text{ m}^3$$

$$\text{Total digester volume} = (2788\text{ m}^3/\text{digester}) \times (2\text{ digesters}) = 5576\text{ m}^3$$

$$\text{Digester feed} = (\text{Primary sludge flow, m}^3) + (\text{Thickened sludge flow, m}^3)$$

$$(194\text{ m}^3/\text{d} + 244\text{ m}^3/\text{d}) = 438\text{ m}^3/\text{d}$$

$$\text{Detention time} = \frac{\text{Digester volume, m}^3}{\text{Digester sludge feed, m}^3/\text{d}}$$

$$\frac{5576\text{ m}^3/\text{d}}{438\text{m}^3/\text{d}} = 12.7\text{ days}$$

In U.S. customary units:

$$\text{Volume} = \pi(\text{radius, ft})^2(\text{height, ft}) + \tfrac{1}{3}\pi(\text{radius, ft})^2(\text{height, ft})$$

$$(3.14)(39.35 \text{ ft})^2(18 \text{ ft}) + \tfrac{1}{3}(3.14)(39.35 \text{ ft})^2(6.6 \text{ ft})$$

$$87\,493 \text{ gal} + 12\,693 \text{ gal} = 98\,186 \text{ gal} \times 7.48 \text{ g/cu ft} = 734\,431 \text{ gal/digester}$$

$$\text{Total digester volume} = (734\,431 \text{ gal/digester}) \times (2 \text{ digesters}) = 1\,468\,862 \text{ gal}$$

$$\text{Digester feed} = (\text{Primary sludge flow, gpd}) + (\text{Thickened sludge flow, gpd})$$
$$(51\,249 \text{ gpd} + 64\,465 \text{ gpd}) = 115\,714 \text{ gpd}$$

$$\text{Detention time} = \frac{\text{Digester volume, gal}}{\text{Digester sludge feed, gpd}}$$

$$\frac{1\,468\,862 \text{ gal}}{115\,714 \text{ gpd}} = 12.7 \text{ days}$$

SOLUTION 30.9

88.4 kg (195 lb)

Determine the caustic dosage required:

$$52 \text{ mg/2 L} = 26 \text{ mg/L}$$

Calculate the amount of caustic required:

In International Standard units:

$$(3400 \text{ m}^3) \times (26 \text{ mg/L}) \times \left(\frac{0.001 \text{ kg/m}^3}{\text{mg/L}} \right) = 88.4 \text{ kg}$$

In U.S. customary units:

$$(0.90 \text{ mil. gal}) \times (26 \text{ mg/L}) \times \left(\frac{8.34 \text{ lb/mil. gal}}{\text{mg/L}} \right) = 195 \text{ lb}$$

SOLUTION 30.10

36%

$$\text{Volatile solids reduction} = \frac{\text{In} - \text{Out}}{\text{In} - (\text{In} \times \text{Out})} \times 100$$

$$\text{Percent volatile solids reduction} = \frac{0.77 - 0.68}{0.77 - (0.77 \times 0.68)} \times 100 = 36\%$$

Based on the volatile solids reduction, the performance of this digester should be reviewed.

Chapter 31
AEROBIC DIGESTION

Problems

PROBLEM 31.1

The primary purpose of aerobic digestion is which of the following?

- **a)** Reduce volatile solids of the sludge
- **b)** Produce a gas that can be used to heat buildings
- **c)** Produce a sludge that is easy to thicken and dewater

PROBLEM 31.2

Which of the following is an advantage of aerobic digestion?

- **a)** Energy reduction
- **b)** Minimal laboratory testing
- **c)** Safe operation (requires less hazardous cleaning and repairing tasks)

PROBLEM 31.3

Operation of an aerobic digester can be improved by which of the following?

- **a)** Using the least concentrated feed sludge
- **b)** Operating as an anoxic reactor
- **c)** Temperature control

PROBLEM 31.4

Which of the following describes the end product of aerobic digestion?

a) Carbon monoxide

b) Ammonia

c) Nondegradable materials such as polysaccharides, hemicellulose, and cellulose

PROBLEM 31.5

Which of the following is a benefit or disadvantage of authothermal thermophilic aerobic digestion?

a) Poor dewatering characteristics of biosolids

b) No odors in the digested sludge

c) No f-Foam control is required

PROBLEM 31.6

Incorporating nitrification and denitrification processes to operating the aerobic digestion system has which of the following benefits?

a) Alkalinity recovery and pH balance

b) Phosphorus is removed along with the nitrogen

c) Improved dewatering

PROBLEM 31.7

Which of the following will reduce odor from the aerobic digestion process?

a) Reducing the organic loading by decreasing the influent sludge mass, total volume added, and the amount of sludge withdrawn from the digester

b) Preventing operation at temperatures lower than 40 °C

c) Adding lime to increase the acid-to-alkalinity ratio

PROBLEM 31.8

Which of the following can control foaming?

a)   Decreasing organic loading
b)   Installing foam-breaking water spray
c)   Adding hypochlorite to the feed sludge

PROBLEM 31.9

Which of the following are components that require regular attention in most aerobic digestion systems?

a)   Aeration or oxygen supply system
b)   Pressure-relief valves
c)   Gas detectors

PROBLEM 31.10

The best aeration device for the aerobic digester will provide which of the following?

a)   Oxygen-transfer efficiencies between 1 and 1.5%
b)   Mixing
c)   Anoxic respiration

PROBLEM 31.11

Which of the following aerobic digestion systems is the least effective in bacterial and viral destruction?

a)   Single-stage completely mixed reactor with continuous feed and withdraw
b)   Multiple tanks in series
c)   Multiple tanks in parallel with no mixing

PROBLEM 31.12

Which of the following is not an accurate statement?

a) Percentage of volatile solids destruction by the aerobic digester is independent of the source of the sludge.

b) If the source of the sludge has a low fraction of digestible organic content, it is difficult to meet the minimum requirements of 38% by the U.S. Environmental Protection Agency.

c) The primary process indicators for monitoring aerobic digesters on a daily basis are temperature, pH, dissolved oxygen, odor, and settling characteristics.

PROBLEM 31.13

What is the thickened solids concentration that is removed from the batch operation or decanting aerobic digester?

a) 1.25 and 1.75%

b) Less than 1%

c) Greater than or equal to 3%

PROBLEM 31.14

Using a gravity thickener in loop with aerobic digestion can produce a digested sludge of what percent solids without using polymer?

a) Less than 2%

b) 3%

c) Greater than 4.5%

PROBLEM 31.15

In aerobic digesters, a typical dissolved oxygen level is which of the following?

a) 0.3 to 3.0 mg/L

b) Greater than 5 mg/L

c) 4.5 mg/L

PROBLEM 31.16

Increased solids retention time of the aerobic digester results in an increase in the degree of volatile solids reduction.

   a) True

   b) False

PROBLEM 31.17

The specific oxygen uptake rate (SOUR) of 1.5 mg oxygen/g solids·h was selected by the U.S. Environmental Protection Agency to indicate that the vector attraction of an aerobically digested sludge by a mesophilic aerobic digester has been adequately reduced.

   a) True

   b) False

PROBLEM 31.18

The operating temperature of the aerobic digester has no effect on the pathogen reduction by the process.

   a) True

   b) False

PROBLEM 31.19

The conventional aerobic digestion process, operated in the aerobic–anoxic mode, has the potential to provide full nitrification and denitrification and a reduction in total nitrogen.

   a) True

   b) False

PROBLEM 31.20

Aerobic digestion of primary sludge results in a higher percentage of volatile solids reduction than aerobic digestion of secondary sludge.

   a) True

   b) False

PROBLEM 31.21

An aerobic digester should always be operated with dissolved oxygen of less than 0.1 mg/L.

    a)   True

    b)   False

PROBLEM 31.22

Aerobic digestion produces a sludge that dewaters easier than raw sludge.

    a)   True

    b)   False

Solutions

SOLUTION 31.1

    a)   Reduce volatile solids of the sludge

SOLUTION 31.2

    c)   Safe operation (requires less hazardous cleaning and repairing tasks)

SOLUTION 31.3

    c)   Temperature control

SOLUTION 31.4

    c)   Nondegradable materials such as polysaccharides, hemicellulose, and cellulose

SOLUTION 31.5

    a)   Poor dewatering characteristics of biosolids

SOLUTION 31.6

a)   Alkalinity recovery and pH balance

SOLUTION 31.7

a)   Reducing the organic loading by decreasing the influent sludge mass, total volume added, and amount of sludge withdrawn from the digester

SOLUTION 31.8

b)   Installing foam-breaking water spray

SOLUTION 31.9

a)   Aeration or oxygen supply system

SOLUTION 31.10

b)   Mixing

SOLUTION 31.11

a)   Single-stage completely mixed reactor with continuous feed and withdraw

SOLUTION 31.12

a)   Percentage of volatile solids destruction by the aerobic digester is independent of the source of the sludge

SOLUTION 31.13

a)   1.25 and 1.75%

SOLUTION 31.14

b)   3%

SOLUTION 31.15

a) 0.3 to 3.0 mg/L

SOLUTION 31.16

a) True

SOLUTION 31.17

a) True

SOLUTION 31.18

b) False

SOLUTION 31.19

a) True

SOLUTION 31.20

b) False

SOLUTION 31.21

b) False

SOLUTION 31.22

b) False

Chapter 32
ADDITIONAL STABILIZATION METHODS

Problems

PROBLEM 32.1

A water resource recovery facility (WRRF) dewaters 9.25 dry metric tons (10.2 dry tons) per day of primary and secondary solids. The dewatering equipment produces dewatered cake with a total solids concentration of 20% and captures 98% of the solids feed. How much dewatered cake is produced per day at this WRRF?

    a)   45.3 wet metric tons (50 wet tons)

    b)   1.9 wet metric tons (2.04 wet tons)

    c)   45.3 wet metric tons (100 wet tons)

PROBLEM 32.2

Name the three basic types of composting.

    a)   Windrow, aerated piles, elongated channel

    b)   Windrow, static piles, in-vessel systems

    c)   Aerated vessels, aerated piles, static windrows

PROBLEM 32.3

At what moisture content will biological activity begin to decrease in a composting system?

    a)   10%

    b)   30%

    c)   40%

PROBLEM 32.4

At what moisture content will the air pore space begin to be blocked in a composting system?

a) 20%

b) 40%

c) 60%

PROBLEM 32.5

A WRRF will compost approximately 45.3 wet metric tons (50 wet tons) per day of dewatered cake at a total solids concentration of 20%. The facility will use wood chips with a 40% moisture content. How many wet tons of wood chips will need to be mixed with the dewatered cake to produce a blended concentration of approximately 40% solids for the compost system feed?

a) 18.12 metric tons woodchip (20 tons)

b) 111 metric tons woodchip (125 tons)

c) 45.3 metric tons woodchip (50 tons)

PROBLEM 32.6

For a lime stabilization process, sufficient lime must be added to do which of the following?

a) Prevent ammonia odors

b) Raise the pH to 12.0 or greater for a minimum of 2 hours

c) Maintain the pH between 7.5 and 8.5 for optimal nitrification

PROBLEM 32.7

Which of the following is true of health and safety considerations for a lime stabilization system?

a) They are less stringent than other processes such as composting, thermal treatment, or anaerobic digestion.

b) They include the use of personal protective equipment such as gloves, safety goggles, and a ventilator.

c) They require continuous monitoring of the atmosphere for combustible gases.

PROBLEM 32.8

What are the typical minimum temperature and pressure of the contents inside a wet air oxidation reactor vessel?

a)   127 °C and 46 600 kPa (260 °F and 6900 psig), respectively

b)   260 °C and 6900 kPa (500 °F and 1000 psig), respectively

c)   500 °C and 20 600 kPa (932 °F and 3000 psig), respectively

PROBLEM 32.9

A WRRF feeds 45.3 wet metric tons (50 wet tons, short) per day of anaerobically digested solids to a heat dryer. The solids have been dewatered to a total solids concentration of 20% before addition to the heat dryer and have a volatile solids content of approximately 55%. How much water must be removed in the heat dryer to produce dried biosolids with a 10% moisture content?

a)   35.3 wet metric tons (39 wet tons)

b)   39.8 wet metric tons (49.4 tons)

c)   159 wet metric tons (175 tons)

PROBLEM 32.10

Multipass thermal dryers produce

a)   pellets that are useable as a fertilizer product.

b)   a very dry, fluffy product.

c)   large clumps of dried sludge.

PROBLEM 32.11

What are the two most-often-used types of incinerators used at WRRFs?

a)   Multiple hearth and static bed

b)   Fluidized bed and single hearth

c)   Multiple hearth and fluidized bed

PROBLEM 32.12

A facility is dewatering a blend of primary and waste activated sludge to a total solids concentration of 16% with vacuum filters using lime and ferric chloride for conditioning. The incinerator is using natural gas fuel at a rate of 1.055 mil. kJ (1.0 mil. BTU) per ton of volatile sludge solids. What can be done to optimize the system to make the incinerator self-sustaining so that fuel does not need to be purchased?

a)   Increase the feed rate and maintain a constant combustion airflow.
b)   Replace the dewatering equipment with another type of unit to increase the incinerator feed concentration to higher than 22% total solids.
c)   Add polymer to the incinerator for conditioning.

Solutions

SOLUTION 32.1

a)   45.3 wet metric tons (50 wet tons)

In International Standard units:

$$(9.25 \text{ dry metric tons}) \times (0.98/0.20) = 45.3 \text{ wet metric tons}$$

In U.S. customary units:

$$(10.2 \text{ dry tons}) \times (0.98/0.20) = 50 \text{ wet tons}$$

SOLUTION 32.2

b)   Windrow, static piles, in-vessel systems

SOLUTION 32.3

c)   40%

SOLUTION 32.4

c)   60%

SOLUTION 32.5

c)   45.3 metric tons (50 tons)

Given

45.3 wet metric tons (50 wet tons) of sludge
1.20% solids in sludge
X = wood chip
40% moisture, which is 60% solids
Final product = 40% solids

Final product

$$\text{Percent Solids} = \frac{(\text{Sludge} \times \text{solids concentration}) + (\text{Woodchip} \times \text{Solids concentration})}{(\text{Sludge} + \text{Woodchip})}$$

In International Standard units:

$$0.4 = \frac{(45.3 \text{ wet metric tons} \times 0.2) + (X \times 0.6)}{(45.3 \text{ wet metric ton} + X)}$$

In U.S. customary units:

$$0.4 = \frac{(50 \text{ tons} \times 0.2) + (X \times 0.6)}{(50 \text{ tons} + X)}$$

$$X = 45.3 \text{ metric tons woodchip}$$

SOLUTION 32.6

b)   Raise the pH to 12.0 or greater for a minimum of 2 hours

SOLUTION 32.7

b)   They include the use of personal protective equipment such as gloves, safety goggles, and a ventilator.

SOLUTION 32.8

b) 260 °C and 6900 kPa (500 °F and 1000 psig), respectively

SOLUTION 32.9

a) 35.3 wet metric tons (39 wet tons)

In International Standard units:

$$45.3 \text{ wet metric tons} \times 0.2 = 9.1 \text{ dry metric tons}$$

$$9.1 \text{ dry metric tons}/0.9 = 10 \text{ wet metric tons of dried product}$$

$$45.3 \text{ wet metric tons} - 10 \text{ wet metric tons of dried product} = 35.3 \\ \text{wet metric tons of water removed}$$

In U.S. customary units:

$$50 \text{ wet tons} \times 0.2 = 10 \text{ dry tons}$$

$$10 \text{ dry tons}/0.9 = 11 \text{ wet tons of dried product}$$

$$50 \text{ wet tons} - 11 \text{ wet tons of dried product} = 39 \text{ wet tons of water removed}$$

SOLUTION 32.10

a) pellets that are useable as a fertilizer product.

SOLUTION 32.11

c) Multiple hearth and fluidized bed

SOLUTION 32.12

b) Replace the dewatering equipment with another type of unit to increase the incinerator feed concentration to higher than 22% total solids.

Chapter 33
DEWATERING

Problems

PROBLEM 33.1

The selection of the proper pump for supplying biosolids to a belt filter press or centrifuge is an important factor to the success of the installation. An ideal pump will produce a steady uniform flowrate with little or no shear. With these parameters in mind, which biosolids pump would not be an appropriate pump for supplying biosolids to mechanical dewatering equipment?

    a) Diaphragm pump
    b) Progressive cavity pump
    c) Rotary lobe pump
    d) Double disc pump

PROBLEM 33.2

Three organic polymers were evaluated for use in a centrifuge dewatering application. The polymer dosage and cost are shown in the table below.

Polymer	Dosage, kg/Mg (lb/ton)	Cost, Polymer $/kg ($/lb)
Polymer A	60 (120)	$0.0382 ($0.084)
Polymer B	5 (10)	$0.423 ($0.931)
Polymer C	3 (6)	$0.723 ($1.590)

According to the performance for each polymer as measured by the cake and percent solids and solids capture efficiency, which polymer has the lowest cost?

    a) Polymer C
    b) Polymer A
    c) Polymer B

PROBLEM 33.3

Biosolids conditioning is a two-step process. In the first step, _____ involves the destabilization of the biosolids particle by decreasing the magnitude of the repulsive electrostatic interactions between the particles. The second step, _____, is the agglomeration of colloidal particles and finely divided suspended matter after the first step by gentle mixing.

PROBLEM 33.4

How biosolids are handled before dewatering will have a significant effect on their dewaterability and the chemical dosage. Which of the following will have the most positive effect on the dewaterability and chemical dosage?

    a)   Produce more primary biosolids by maximizing suspended solids removal.

    b)   Store biosolids in an aerated tank before processing.

    c)   Operate anaerobic digesters with sufficient detention time to meet volatile solids reduction requirements for Class B biosolids.

    d)   Process alum sludge from the water treatment facility at the water resource recovery facility.

PROBLEM 33.5

An operator needs to remove biosolids from an aerobic digester and dewater the biosolids using the facility's sand drying bed. The sand drying bed is 20 m × 30 m (65.6 ft × 98.4 ft) and the sludge depth is 0.152 m (6 in.). How many cubic meters (gallons) of aerobically digested biosolids will the operator be able to apply to the sand drying bed?

PROBLEM 33.6

The feed biosolids concentration to a mechanical dewatering process such as a belt filter press or centrifuge is an important process variable. Which description best represents the effect of a dilute feed biosolids on the performance of mechanical sludge dewatering equipment as measured by polymer dosage, cake solids, and throughput?

    a)   Polymer dosage, cake solids, and throughput increase

    b)   Polymer dosage decreases, cake solids and throughput increase

    c)   Polymer dosage increases, cake solids and throughput decrease

    d)   Polymer dosage, cake solids, and throughput decrease

PROBLEM 33.7

An operator is dewatering anaerobically digested biosolids on a 1-m belt filter press. The anaerobically digested biosolids are a blend of primary and waste activated biosolids. The feed sludge concentration is 2.5% solids and the digested biosolids flowrate is 13.63 m³/h (60 gpm). What is the solids loading rate (in kilograms per hour [pounds per hour])?

PROBLEM 33.8

An operator determines that the belt filter press polymer dosage is higher than normal. Which of the following causes would explain the higher-than-normal polymer dosage?

    a) Polymer injection ring hoses plugged

    b) Low feed biosolids concentration

    c) Biosolids/polymer mixing device is in need of adjustment

    d) All of the above

PROBLEM 33.9

Given the following data, determine the solids capture efficiency (%) for a centrifuge. The feed sludge concentration is 3.0% solids, the cake is 25% solids, and the filtrate solids concentration is 0.15% solids.

PROBLEM 33.10

An operator typically makes up an emulsion polymer solution at 2.0% by volume. The polymer vendor recommends that the operator make up the polymer solution at 0.75% by volume. The polymer mix tank volume is 4540 L (1200 gal). How much polymer is required to make up the 0.75% concentration?

PROBLEM 33.11

In reviewing the recent operating logs for the centrifuge, the operator finds that the overall performance has dropped in the last week. Which of the following conditions is not a possible cause for the reduction in overall performance?

    a) Sludge no longer reacts well with the polymer

    b) Feed solids are too thick

    c) Feed rate is too high

    d) Conveyor bearing is failing

PROBLEM 33.12

Effective polymer evaluation and selection is critical to maintaining cost-effective biosolids dewatering and minimizing the potential for future polymer-associated performance problems. From the following list of suggestions, determine the suggestion that will likely create future performance issues for the selected polymer?

a) Establish proper performance criteria.
b) Standardize sludge conditions.
c) Run product evaluations over a 1- to 2-month period.
d) Use standard sampling and analytical methods.

PROBLEM 33.13

A water resource recovery facility uses a filter press for anaerobically digested biosolids dewatering. Each filter press cycle will process a 1 metric ton (2205 lb) of biosolids and conditioning chemicals. The typical ferric chloride and lime dosages are 50 kg/metric ton (100 lb/ton) and 250 kg/metric ton (500 lb/ton), respectively. If the facility must dewater 3.5 metric tons/d (7718 lb/d) biosolids, how many filter press runs are required per day?

PROBLEM 33.14

A process engineer finds that the polymer solution feed pump feed is frequently tripping on high-discharge pressure. In reviewing the dry polymer make-down system, it is determined that the polymer is being made up at 1.0% concentration. What is the optimum range for make-up of a dry polymer solution?

a) 0.5 to 2.0%
b) 0.10 to 0.50%
c) 5.0 to 10.0%
d) None of the above

Solutions

SOLUTION 33.1

a) Diaphragm pump

A diaphragm pump will create pulsations in the flow stream, thereby adversely affecting polymer application, which is at a constant flowrate.

SOLUTION 33.2

c)   Polymer B

In International Standard units:

$$\text{Polymer A} = 60 \text{ kg/Mg} \times \$0.0382/\text{kg} = \$2.292/\text{Mg}$$
$$\text{Polymer B} = 5 \text{ kg/Mg} \times \$0.423/\text{kg} = \$2.115/\text{Mg}$$
$$\text{Polymer C} = 3 \text{ kg/Mg} \times \$0.723/\text{kg} = \$2.169/\text{Mg}$$

In U.S. customary units:

$$\text{Polymer A} = 120 \text{ lb/ton} \times \$0.084/\text{lb} = \$10.08/\text{ton}$$
$$\text{Polymer B} = 10 \text{ lb/ton} \times \$0.931/\text{lb} = \$9.31/\text{ton}$$
$$\text{Polymer C} = 6 \text{ lb/ton} \times \$1.590 = \$9.54/\text{ton}$$

SOLUTION 33.3

Biosolids conditioning is a two-step process. In the first step, <u>coagulation</u> involves the destabilization of the biosolids particle by decreasing the magnitude of the repulsive electrostatic interactions between the particles. The second step, <u>flocculation</u>, is the agglomeration of colloidal particles and finely divided suspended matter after the first step by gentle mixing.

SOLUTION 33.4

a)   Produce more primary biosolids by maximizing suspended solids removal.

Primary biosolids are the easiest to dewater and require the least amount of polymer. The other options listed will tend to increase the polymer dosage and decrease biosolids dewaterability.

SOLUTION 33.5

91.2 m³ (24 142 gal)

In International Standard units:

$$20 \text{ m} \times 30 \text{ m} \times 0.152 \text{ m} = 91.2 \text{ m}^3$$

In U.S. customary units:

$$65.6 \text{ ft} \times 98.4 \text{ ft} \times 0.5 \text{ ft} \times 7.48 \text{ gal/cu ft} = 24\,142 \text{ gal}$$

SOLUTION 33.6

c) Polymer dosage increases, cake solids, and throughput decrease

SOLUTION 33.7

341 kg/h (750 lb/hr)

In International Standard units:

$$13.63 \text{ m}^3/\text{h} \times 25\,000 \text{ g/m}^{3}* \times 1 \text{ kg}/1000 \text{ g} = 341 \text{ kg/h}$$

$$*2.5\% = 25\,000 \text{ mg/L and mg/L} = \text{g/m}^3, \text{ so } 2.5\% = 25\,000 \text{ g/m}^3$$

In U.S. customary units:

$$(60 \text{ gpm} \times 60 \text{ min/hr}) \times 0.025 \times 8.34 \text{ lb/gal} = 750 \text{ lb/hr}$$

SOLUTION 33.8

d) All of the above

For a), the polymer will not be distributed evenly to the sludge flow stream. For b), a drop in the feed sludge concentration will increase the polymer dosage. For c), overmixing or undermixing of the sludge and polymer will increase the polymer dosage.

SOLUTION 33.9

95.5%

$$\text{Capture erfficiency} = 100\% \times \left[\frac{\text{Cake solids (Feed} - \text{Filtrate)}}{\text{Feed solids (Cake} - \text{Filtrate)}} \right]$$

$$100\% \times \left[\frac{25 \times (3 - 0.15)}{3 \times (25 - 0.15)} \right] = 100\% \times \left[\frac{71.25}{74.55} \right] = 95.5\%$$

A capture efficiency of 95.5% is an acceptable value.

SOLUTION 33.10

34 L (9 gal)

In International Standard units:

$$\text{Polymer, L} = 0.75/100 \times 4540 \text{ L} = 34 \text{ L polymer}$$

In U.S. customary units:

$$\text{Polymer, gal} = 0.75/100 \times 1200 \text{ gal} = 9 \text{ gal polymer}$$

An emulsion polymer is typically made up between 0.1 and 1.0% by volume. Polymer concentrations greater than 1.0% by volume may be difficult or impossible to pump because of the high viscosity.

SOLUTION 33.11

d) Conveyor bearing is failing

A conveyor bearing that is failing will cause a gradual increase in the centrifuge vibration. This problem will affect the ability to operate the unit only.

SOLUTION 33.12

c) Run product evaluations over a 1- to 2-month period.

The product evaluation should be run over as short a time period as possible—1 week.

SOLUTION 33.13

Five runs per day

In International Standard units:

$$\text{Runs/d} = \frac{3.5 \text{ Mg/d (1000 kg/Mg)}}{1000 \text{ kg} - 50 \text{ kg} - 250 \text{ kg}} = 5 \text{ runs/d}$$

In U.S. customary units:

$$\text{Runs/d} = \frac{7718 \text{ lb/d}}{2205 \text{ lb} - 100 \text{ lb} - 500 \text{ lb}} = 5 \text{ runs/d}$$

SOLUTION 33.14

b)   0.10 to 0.50%

Polymer pumps can trip out and the polymer dosage can significantly increase—polymer will not mix very well with the waste activated sludge. Decreasing the polymer concentration to 0.3% can resolve the problem and bring the polymer dosage back into the normal range.

CPSIA information can be obtained
at www.ICGtesting.com
Printed in the USA
FSHW010911310519
58558FS